Brothers & Sisters
Like These
Volume III

An Anthology of Writing by Veterans

Brothers & Sisters Like These
Volume III

An Anthology of Writing by Veterans

Brothers & Sisters Like These
Copyright © 2025 The NC Veterans Writing Alliance

All rights reserved. No part of this publication may be reproduced, distributed, or transmitted in any form or by any means, including photocopying, recording, or other electronic or mechanical methods, without the prior written permission of the publisher, except in the case of brief quotations embodied in critical reviews and certain other noncommercial uses permitted by copyright law. For permission requests, write to the publisher, addressed "Attention: Permissions Coordinator," at the address below.

ISBN: 978-1-959346-90-6 (Paperback)
Library of Congress Control Number: 2025936204

Cover Design: Ashley Minnick
Book Design: Erin Mann

Printed in the United States of America.
First printing 2025.

Redhawk Publications
The Catawba Valley Community College Press
2550 Hwy 70 SE
Hickory, NC 28602
https://redhawkpublications.com

Table of Contents

7	Ron Toler	Introduction
9	Richard Epstein	"Night"
10	Wallace Bohanan	"Unsent Letter to Family"
11	Dorian Dula	"Why"
13	Mike Smith	"One Hundred Puppies"
19	Monica Blankenship	"Survivor's Guilt"
20	John T. Hoffman	"The Last DROS"
22	William "Pete" Ramsey	"RATED Mature Audience ONLY"
26	Alan Brett	"Memories"
28	Dean A. Little	"The Mission"
31	R. Kevin Wierman	"That Guy: Squirrel Murphy USN - BUSC/SCW"
33	Steve "Buck" Owens	"The Worst and Best Day of My Life"
38	Donna Culp	"When You Really Have a Team"
39	Michael D. Hebert	"Terrorists"
41	Richard Epstein	"What Can Happen?"
42	Tom Baker	"First Time Outside the Wire"
45	John Mason	"A Life Spent"
47	Larry Kipp	"What I Learned from the Vietnam War"
50	John T. Hoffman	"Our F Troop, 8th Cav 'Angel'"
55	Carl T. Zipperer	"Ode to Donald"
57	Sarah Scully	"Tent City"
59	Wallace Bohanan	"Home"
60	Michael D. Hebert	"The Cold War"
62	Richard Epstein	"Eyes that Talk"
63	Tom Hickey	"Arrival at an Anti-Communist Guerilla Camp"
65	Alan Brett	"Fond Memories"
66	Harold (Ted) Minnick	"Boom"
69	Allan Perkal	"Dear Momma: Letter Home from Nam"
71	Sarah Scully	"Letter to the Enemy"

74	Ron Toler	"What I Carried"
75	Beth Angel	"Transition to Limboland"
76	Allen Utterback	"Gambit Prayer"
77	Roy Moore	"Survivor's Guilt"
78	Jackie White	"Leaving El Toro"
80	Wallace Bohanan	"Letter to That Artillery Crew"
81	Beth Angel	"You Saw Me but You Didn't"
82	Allan Perkal	"Who am I"
83	Stephen Henderson	"Reflection"
84	Roy Moore	"What Saves Us"
85	Ron Kuebler	"Veterans Day USA"
86	Marsha Lee Baker	"Wife of a Vietnam Veteran: Married to the War"
90	Richard Epstein	"4th of July: Hot off the Grill"
92	Donna Culp	"Letter to the Enemy: I Haven't Thought About You a Long Time"
93	Emiliano Enea	"Where I'm From"
95	Allen Utterback	"What Saves Us"
96	Charles R. Duke	"Meat and Three"
98	Roy Moore	"Letter to my Enemy"
99	Ron Kuebler	"Suicide: What is it Like"
100	Donna Culp	"When I Knew I Was Home"
102	Stephen Henderson	"Memorial Day, 2019"
105	Dorian Dula	"Survivor Guilt"
107	Beth Angel	"Standing in Front of You"
110	Tom Hickey	"Eight Green Thumbtacks"
112	Ron Kuebler	"Farm that Heals"
113	Charles R. Duke	"Aftermath"
116	Ted Minnick	"What is a Veteran"
118	Carl T. Zipperer	"Home: Back in the World"
120	Allen Utterback	"The Veterans Healing Farm"
122	Gerry Nieters	"Death by PTSD"
125	Larry Boggs	"My Talisman"
127	Glossary of Terms	
131	Contributors	

Introduction

Welcome to the third book in our series of Veteran stories. In the past, we shared with you our beginnings, our history, and some of our processes, but the underlying story is that of the Veteran. Veterans comprise a small portion of the general population. Currently, less than 1% of the population serves in the military. Military service is a unique way of life, and each branch is unique in its training, language, and mission. Yet there is a common bond: they all swore an oath to serve and protect. They swore an oath to make the ultimate sacrifice if needed in the service of our country and our ideals.

We have been fortunate to have a diverse group of Veterans join us in sharing their stories. People from all backgrounds have stepped up to write their stories, to share their experiences, and to once again become part of the tribe. They all had different reasons to do so, whether they served for a short period or as long as 40 years. They may have volunteered or been drafted. Many served in combat, and others supported those combat veterans, both in the war zone and at home. Some served in the jungles of Vietnam, while others served in the deserts of Kuwait, Iraq, or Afghanistan. Some got a "Hero's Welcome Home," and some were reviled upon their return. Many may have Post-Traumatic Stress or a Traumatic Brain Injury, and many suffer Survivor's Guilt, yet they are all Brothers and Sisters in arms forever, united by the oath they swore.

All these stories are part of their history, but they are also part of your history, of our nation's history. What you hold in your hand are a few of the stories written and shared in the classroom. An emphasis has been put on what they wanted to share with you to create an understanding of what they carry. These are stories of bravery, sacrifice, lost innocence, lost youth, lost comrades and friends, but they are also stories of healing. Our goal is to share these stories in the hope that they can assist in the healing process.

We have used various avenues to share these stories in the past, including staged readings at community theaters, playhouses, university campuses, libraries, in the Court system, and at the Veterans Healing Farm. We have read for Memorial Day and Veterans Day Ceremonies at Veterans Cemeteries and in city council chambers. Black Box Dancers, a modern dance company, allowed us to read and dance with them while filming a documentary. All these endeavors are designed to share these stories with the public while allowing the healing process to occur for both the veterans and the individuals.

The North Carolina Veterans Writing Alliance, known as Brothers and Sisters Like These, is expanding this program to reach as many Veterans in as many locales as possible. Zoom classes have allowed international participation. I have had the privilege of hearing these stories firsthand and am now happy to share them with you—the stories you will read have been buried for decades, some for over half a century. Three generations of Veterans have bonded while sharing their stories. By reading this anthology, we believe you will gain insight into a small part of their journey to dark places and a part of their journey back into the sun.

Ron Toler

USAF Pilot 1970-1977 /Vietnam Veteran

Night
Richard Epstein

Sixteen dollars and forty-eight cents
    in   my pocket
Drove across Texas   had to siphon gas
   along the way

Stuck in New Orleans
    Pumped gas   sold used tires
from a car lot   next door.

Hungry stuck on the road again
    Storm clouds   gather
Draft board   won't wait
    Guard duty moonlit night

Trees   bushes   move
   My buddy tells me: blink
Don't   stare   look away
   then back

This mountain   holds onto us
   This mountain   may bury us
Trip flares
   They're coming in

Ready left!   Ready Right!

Bugles   blare   whistles shriek
    the earth throws-up its dirt
artillery   mortars tracers right

Johnson's hit
   It's another   firefight

Unsent Letter to Family
Wallace Bohanan

Hi Everybody,

    It's just another day on the DMZ. I've gotten used to the suffocating heat. Yet, mortars, rockets, and artillery screaming in every day make it difficult to take a shower. I don't change clothes that often anyway, and I stopped wearing underwear long ago. We eat C-Rations for breakfast, lunch, and dinner, along with water and the occasional warm beer. We alternate 24 hour shifts on the gun, but we never have a day off. I've forgotten what a weekend feels like. I'm either humping 178lb artillery rounds so we can fire downrange, or digging dirt, or unloading trucks, or....

    Don't get me wrong. I'm not complaining, just explaining. Believe it or not the heat is better than the monsoon season, when it pours down rain every day. Being soaked to the bone has taken on a new meaning for me. Wearing clothes that never dry because you are constantly exposed to the downpours has brought its own deep level of depression. And the mud! The wet, sloggy, slippery, sticky mud makes every movement slow and precarious. Morale is really low during the monsoons.

    We pretty much live underground in our bunkers these days. Tell Carol thanks for the cookies and tell dad thanks for the smokes. Everyone in my bunker appreciates them. The only downside is that these goodies remind us of home. Consistently dealing with incoming rounds, fire missions, blown-up equipment, and combat casualties is taking its toll on me. Longing for home just makes it more difficult to stay mentally intact.

    Please know that all of us here are prepared to make the sacrifices that war demands of us. Yet, please send more cookies.

Your Loving Son,
Wally

Why
Dorian Dula

Do you ever wonder why you make the decisions that you do? I think about some of my previous decisions a lot and wonder why. Probably the first big decision of my life was to join the Marine Corps. I could have stayed in college and got a deferment and avoided the war possibly. But I wasn't ready for college and I don't want to beat a dead horse as I've discussed this before, but my mother was a mess, and I found out my girlfriend—who I worshiped–was having sex with some other guy. Welcome to the world of being in love.

Anyway, I was 19 and trying to figure out what was going on in my life. Did I join the Marines to test myself to see if I could handle it? Was it a sense of patriotism for my country? I came from a patriotic family. My paternal grandfather was a Captain in the Army in WWI. My father was in the Navy in the South Pacific in WWII. And his ship was in the battle for Leyte Gulf. His brother my uncle was a fighter pilot in WWII and was shot down behind enemy lines in Burma and, although wounded, he made it out.

I think it was probably a combination of those three things.

1. Wanting to get away from that small farming town I lived in

2. Testing myself

3. Patriotism for my country

At that time, the Marine Corps had a 2-year enlistment. It was 2 years active duty, 2 years active reserves and 2 years inactive reserve. You were just on some list. But only joining for 2 years you were 99% sure you were going straight to the infantry. Which I did.

But after ITR (Infantry Training Regiment) I was in Staging Battalion, ready to go to Vietnam. A week or so before I was to leave, I got sidetracked and they sent me to the 3rd MP Battalion. We trained and went over to Vietnam on a Navy Troop Carried. In Vietnam I was initially a Prison Guard at a POW Compound in DaNang. After a month of this I guess I was Gung Ho and wanted to really be in the war and not "In the Rear with the Gear." A Gunnery Sergeant came to our compound once and asked for volunteers to come out to the "Line Companies" which were the Infantry Battalions out in the Bush.

Why did I do that? I could have stayed relatively safe in DaNang, pretty much out of harm's way. Was it because I didn't join the Marine Corps to be a Prison Guard? Did I do this to prove something in myself? To this day I wonder why.

So, I wind up in Charlie Co., 1st Battalion of the 5th Marine Regiment, of the 1st Marine Division. This was late April 1967. My first firefight was a few days later. This was followed by platoon and company sized sweeps and operations.

One night on bunker watch, I shared a bunker with a marine who worked in the company office. An office pogue, as we called them. We began talking, and as usual on bunker watch, you find out everybody's life history in about 30 minutes. Somehow the subject came up that I had attended a Community College for a year. He asked if I could type, and I said yes I could. He said one of the other marines that worked in the office would be rotating home soon and they needed a replacement. He asked if I was interested. I had been in the bush for 3 months already and I decided to do that. So, I became an office pogue.

After a month or two I felt guilty that I had abandoned my buddies in the 2nd platoon. They were beating the bush and I once again was "in the rear with the gear." Not as safe as when I was in DaNang guarding prisoners as we were still in the Que Son Valley, the home of the 2nd NVA Division. So, I went to the 1st Sgt. and told him I wanted to go back to the 2nd platoon. Again, why? I could have skated through my whole tour as an office pogue. I guess I wasn't meant to be a pogue.

After several more operations including Operation Swift, where we lost 125 marines, and later when we went to Hue City during the Tet Offensive of 1968, and 216 Americans died and I took a bullet, I questioned my decisions.

But looking back I wouldn't have changed a thing and would do it all over again.

One Hundred Puppies
Mike Smith

I stopped at the Seven-Eleven to pick up a pack of cigarettes on my way. One man stood in front of me at the register. He slammed his hand on the counter.

"It's too much! How can you thieves charge that for bread, for Christ's sake?"

"I can put it back if you don't want it, sir," the cashier said.

The man turned to me and opened his mouth.

I spoke. "DON'T.

…Ask.

Me."

I looked into his eyes, through them and beyond them. I smelled the tangy, soft, but instant fear come on him. I didn't take my eyes away. "Pay or get the fuck out of my way."

The man forgot where he was. I watched him deal with his terror until he remembered he had come to buy a loaf of bread. His hands opened toward me at his waist, palms out; then he turned back to the counter, his right-hand stumbling until it found his wallet. He grabbed his bread and change, then left, head down, avoiding eye contact with everyone.

I stepped to the counter, laid my money down, and asked for a pack.

The cashier made change, and I said "thanks."

"Thanks to you, too," said the cashier.

I shook my head no, turned and walked out, wondering who the hell I was. Bread man was still sitting in his car, looking as if he was trying to figure out how buying bread had become a life-and death situation. I started my car and drove down Mason Avenue toward the Seabreeze Bridge, mystified at the massive anger in me over a stupid loaf of bread.

Sarah called, "Come in!" when I knocked. Inside, the house on North Halifax Drive was almost completely bare; just a sofa in the middle of the living room sitting on a bare wooden floor. There were no pictures on the wall. Sarah sat on the sofa, knitting and fully pregnant.

"Mike! How are you? Freddy told me you went into the Navy! Tell me! What have you been doing? Roger'll be back soon, can you wait?"

"I guess so. I'm home on leave."

"So, tell me; where have you been?"

"Hawaii, the Philippines and stuff."

Sarah, slender as always, put the knitting needles down on her incongruously huge tummy and looked at me. I sat down on the arm of the sofa.

"Vietnam?" she asked.

"Yeah."

"Oh."

With Sarah's gaze on me, I dropped my eyes to the floor, then looked around the room.

"This is Roger's dad's place, isn't it?" I said.

"Yeah. He built a new house up in Ormond Beach and let us rent this until Roger goes to school."

"Where's he goin'?" I asked.

"He's starting at the University of Georgia engineering school in September. Right now, he's working as an electrician on new houses."

"Great!"

Roger came in, dropping tools at the door. He was the same Roger, unchanged since the last time I had seen him. His energy and smile jumped into the room ahead of him. I could see the old humor behind his eyes.

"Mike! Where have you been?" he grabbed my hand and shook it, then plopped on the sofa between Sarah and me.

"Mike's in the Navy. He's on leave from Vietnam."

"No kidding. Didja see any action?"

"Not much." I looked at the floor again.

Silence grew around me, uncomfortable, but more and more familiar.

"What's it like, Mike?" Sarah asked.

I thought a minute. "Remember when your dog had puppies and you cried because one of them was stillborn? …Kind of like that."

Sarah and Roger looked at each other, then back at me. "Things are just different," I said.

The humor behind Roger's eyes disappeared. "Did ya hear what happened up at Kent State? The National Guard just opened up on the students and killed a bunch of 'em!"

"Yeah, I heard… those guys musta been really scared."

"I bet so!" Roger said, "The Guardsmen were just mowing them down!"

I couldn't answer. I shouldn't be here; Roger thought I meant the students. I couldn't tell him what it feels like to see your own death coming at you; how small your weapon seems; the level of terror it would have taken to begin firing.

"Yeah, I guess so…"

Roger and Sarah exchanged another look.

Roger offered, "Hey, did you know Jack Dillard is in town? He's on leave from the Air Force."

"No kidding? Where's he stayin'?"

"Same house, up with his folks on A1A in Ormond."

"Wow, thanks! I gotta swing up there and say hi."

"Yeah, you oughtta do that…"

I stood up and said something for a goodbye. I wished Sarah well with the new baby, and Roger shook my hand at the door and told me to come back when I got leave again. I knew I wouldn't be back.

I drove slowly north on Halifax, aware of bright sunshine, the ocean breeze, and how seriously warped Daytona had become for me. I thought about National Guardsmen surrounded by hundreds of students, finally overcome by fear.

Jack's parents' house was just north of Granada Boulevard on the beach side of Highway A1A. Mrs. Dillard answered the bell.

"Mike! How are you?"

"Hi, Mrs. Dillard."

"Please, wait here and I'll go get Jack. He'll be so glad to see you!"

I walked out on the back deck overlooking the beach, checking the surf for rideable waves, wondering at such an old habit coming back to me.

Jack and his mother came out on the deck.

"Hey, Jack. Roger and Sarah told me you were in town. How's it goin'?"

"Okay."

"You boys just sit right here and I'll get you some iced tea."

"Thanks, Mrs. Dillard."

Jack and I sat down at the round deck table together. The sun was low enough to give us shade from the house by now. We left the table's big umbrella furled; both gazing out over the Atlantic Ocean.

"Roger and Sarah said you're in the Air Force."

"Yeah. How 'bout you?"

"Navy."

"Really? You on a ship?"

"Yeah, mine sweeper. What about you?"

"I got into the Air Force Academy. I'm a pilot now."

I looked at Jack. I saw his eyes. "You've done a tour, haven't you?"

"Yeah."

We looked back at the ocean, both of us, sitting in the afternoon shade, feeling the breeze.

"I just volunteered to go back," Jack said.

"No, don't, Jack!"

"I have to do it."

"Look, man, I already have my orders. I'll be shipping out as soon as this leave is over. You stay here… I'll go for both of us."

"I have to, Mike… I just have to."

"Freddy said you're married, right?"

"Yes. Barbara and I got married last year."

"Okay, Jack, there's your reason. Stay here. Stay with Barbara and make a baby like Roger and Sarah. I swear, I'll go for both of us."

"I don't know these people anymore, Mike."

I hung my head at that. "Yeah. Sarah asked what it was like, and I tried to say it was like when her puppy died. I don't think she got the message."

"None of them do."

Jack's Mother came out with two tall glasses of tea.

"What are you two boys talking about?"

I looked up at Mrs. Dillard as she set the glasses on the table, not knowing what to say.

"One hundred puppies, Mom. We're talking about one hundred puppies."

"Well, have fun talking…"

Mrs. Dillard went back into the house.

"Mike, if they said you didn't have to go, would you stay Stateside?"

"No, I'd go… but I have an excuse. We have a bunch of cherries on board who don't know their asses from holes in the ground."

"That's my point."

"Hey, Jack?"

"Yeah?"

"Fuck you for bein' right."

Jack snorted a laugh. We both picked up our glasses.

"Here's to one hundred puppies," I said.

"To one hundred puppies."

We clicked the glasses together and drank the Southern sweet tea, watching the ocean as the afternoon sun dropped low, bringing a chill into the onshore breeze. Jack told me Barbara was in Gainesville at the University of Florida; we talked some about what other old friends had done. We sat comfortably together, two old men in young men's bodies whose mothers still called them 'boys.' Jack saw me out the front door when I left at sundown. As I walked toward my car, he called from the porch.

"Hey, Mike!"

"Yeah!"

" …Take care of the puppies, Sailor!"

"…Roger that. You, too."

We stood in the evening light facing each other, two friends in a town full of strangers.

"Jack?"

"Yeah?"

" …Fly safe, Wing-nut!"

I snapped a salute. Jack returned it.

Daytona Beach was still strange to me as I drove home on Riverside Drive, watching the lights across the river flicker on the water. I had come to this place hoping to see my high school friends; I had found one. Only Jack and I knew that our true home was thousands of miles away. In a place where fear is a daily habit and much of the local population wanted us dead, we had found brotherhood.

Survivor's Guilt
Monica Blankenship

"Why am I still alive?"

It was unspoken, this question, most of the time. But evident, even still, in the eyes of so many of you as patients.

I only see the aftermath, and by 1975 you are already burying it, with so much else, deep in your soul. Jumbled in with being disrespected, loss, futility, and for some, self-loathing.

In the ICU you are my age; young, virile, handsome, strong.... yet every pore oozes alcohol and despair....one more overdose of booze and drugs....trying to bury what you can't deal with…comatose, intubated, on a ventilator.

I want to save you. So, I bathe you, turn you, rub legs and arms, suction you, check your vitals and lab results, poke you with needles, flood you with IV fluid and meds, draw your blood. Talk to you. And pray.

But you die.

I should have been able to save you for another day, another try to make your hurt go away. Just like your nightmare about your buddy from over there, I couldn't save you.

I am sorry.

And guilty. I should have been able to make this better for you. Tried, yes, but failed. Maybe on all of us, this, but right now on me.

The Last DROS
By John T. Hoffman

Both in Flight School and "down range," your "stick buddy" shared every flight, every mission, you shared your "C"s and your beer, you shared every near miss and every one that wasn't. As you moved into combat operations you gained a few, a very few, more stick buddies, with whom you grew just as close. You covered each other's back, no matter what happened. You and your stick buddies waited out your yearlong tour of duty in the heat, dampness, evil smells and with that constant companion, death. You counted down the days until your "Date Return from Overseas Duty" or DROS. Some used small scratches on your helmet, some tied a small knot for each pacing week on your pace cord or on a small piece of green paracord on your flight helmet bag, if you were an Army Helicopter Pilot, as I was.

But some of our stick buddies were not so lucky. That "big sky... little bullet" theory was not always to your advantage. Sometimes one or more of our stick buddies occupied the same small space in the sky as that enemy bullet...or missile...or anti-aircraft-gun fired airburst. In most cases, but not all, their DROS was suddenly at hand. Perhaps on a stretcher, if they were very lucky...or in a body bag. For some, this DROS took decades. For Dusty, it was nearly four.

If you were successful surviving all of the routine deadly events that were every day in South Vietnam, and you managed to stay alive dodging the "green balls" of fire from enemy anti-aircraft weapons, or the SA-7 missiles for a full year, you reached your own DROS, based upon the calendar, and not enemy action. But even after returning home to family and the land with real ice cream, good beer, baseball and so many other of life's joys, your time in that green hell was never far from you. Most of us stayed in touch with our stick buddy. You were each other's crutch as you dealt with the aftermath of that long year, or more, over there. But Dusty was not there.

So, as your life moved on, as time ticked by, and as life threw curve balls, you and your surviving stick buddies were there to support each other and what was precious to you both. And as the challenges of age set in, that support gets harder and harder to render. Distances seem greater, travel gets harder and the energy to move gets less and less each day.

Ultimately, there is the inevitable parting on that final mission, that each stick buddy must make alone. But each knew and know the others are there for them, for their family and for that big DROS in the sky that lays "down range" for all of us. As each of us pull pitch for the last time and you "slip those surly bonds of earth" on your final mission, know that all your surviving "Stick Buddy Brothers" are here for you and your loved ones.

Fly safe Dusty, and welcome home!

George "Dusty" Holm, Captain, US Army, US Army Flight School Class 71-30, KIA June 12, 1972.

RATED Mature Audience ONLY
William "Pete" Ramsey

About the only way to describe it adequately is to say it is just a place of bestiality. Of all the times we have been here, the only thing we have taken away at the end of the day is more and more sorrow. Our interactions within this area are the kind that only seems to harden you inside and leave a stench in your soul. Even fear is facing a difficult challenge in piercing your psyche. It is just the kind of place the enemy likes to screw with you and, God, do you hate him for doing that.

Monsoon time has left and now the earth is becoming hard as concrete. The grassland plain itself is covered by smaller trees and scraggly palms maybe fifteen feet tall tops. Clumps of tall grasses and smaller bushes gather in thickets but give way to open areas on occasion. Sightlines are blocked and moving through the space either on the track or on foot creates a maze-like experience. Human nature deposes you to seek the thinnest barriers to breach. Consequently, those passage points are where lethality reigns.

The Gooks, Chuck, Victor Charlie, sow their choicest surprises along those boundaries. Our own defective ordinance is the main source of weaponry. They then retreat to watch the fireworks and leave us nothing to direct retaliation towards. All we get to do is clean up the grisly results which then leads to our own barbarity at times. Whoa, not supposed to say that lest it gets labeled as hatred, warmongering. So, let's just think of it as cataclysmic frustration.

The result of firefights mixed with our own use of explosives rends their bodies into beyond grotesque junctures and dispositions. I notice we no longer look away from the results of our actions. An ugly tally sheet is being kept in our minds. It is sickening and fulfilling. Wait! Not supposed to say that. Hold it inside. Remember your Boy Scouts Oath. God and Country. We are the good guys, remember? Yet drinking from that Challis in the past is what helped put you here to begin with. Its contents are corrosive to the soul and will prove to be a lifetime malignancy.

I have been given the chore of the care and feeding of a Chieu-Hoi or enemy informant. He got wounded and captured, and rather than fall on his sword, decided to assist his enemy. A noble thought, but maybe only a ruse to get healthy and fatten up before returning to his friends back here in this hellhole. Then more misery can be inflicted upon us.

Oddly, I find nothing but ambivalence in my task. No fear, no righteous indignation, no seething hatred. He is the stereotypical image of a VC replete with black pajamas and Ho Chi Minh sandals. I should hold some angst and wariness surrounding my chore, yet indifference holds the upper hand. Have I become so jaded that even my own possible death has lost its grip, or have I gone mad?

The night before, when he arrived, I stared at him closely, watching his eyes for even the slightest tell. I did not see the hatred coming from his eyes as is so often directed at me from even the children and mama-sans I have encountered here thus far. I put him near me and made sure no weapons were within reach and that both of us were in sight of whoever is on guard duty.

I wonder if he is like me, another soul caught up in this nightmarish moment? I have long since given up any pretense of believing in the outright lies which have put me at this place and time. It is my own g-d fault. Maybe I secretly believe I should die for such a grievous error. Maybe the shame seeks balance. Yeah, I am crazy, but so what?

The morning comes and my charge is still here. We are both alive as well. His info is being reviewed and in a bit, a team of Vietnamese and American officers/advisors will go out to verify its validity. The info he provided must be valuable because we have a MACV Colonel with his own radioman, a Captain, no less.

Released of my charge I put my camera to work recording faces of all involved in the day's action. We will sit here until called forward or something occurs. It is a very short wait.

They have only moved a short distance when the quietness is suddenly shattered by an explosion and then silence. I swing the camera to a rising cloud of black smoke. Word soon circulates that the American Captain has hit a tripwire.

A Medi-vac is called for to collect the remains, a single shin bone. As the chopper pulls away, a crewman flashes me the 'peace sign' and is recorded on an Ektachrome roll of film. It bears the image of the Captain with a morose look on his face, staring at the ground. Did he know what was coming? You sometimes do, you know. Along with that image, is one of the POW with that still-expressionless face. Another day begins.

It is years later now. Many stories have played out about the Vietnam Veteran in books, plays, movies. Recently, an exposition on the closing days of Vincent Van Gogh has been crossing the Nation. Basically, the imagery displayed portrays his early arrival in Provence where he first captures all the brightness of landscapes, flowers, even the people of the countryside. All are cast in a profuse glowing display. Accompanying music fills your ears and mind with bright themes, uplifting your mood. The room is filled top to bottom, floors and ceiling, and all around encapsulating your senses in a psychedelic barrage. The whole atmosphere is charged with hope and peacefulness.

But slowly a darker and turbulent imagery begins creeping into the works. Along with the sights, the music shifts as well, now darkness and worrisome displays are laying claim to Van Gogh's beautiful expression on the screen. And it is at that moment in the show which Samuel Barber's Adagio for Strings begins playing in the background. The chords coming from the strings create reverberations which psychically resonate in your body. The choreographer has found a piece of music that haunts me and will always haunt me.

In the movie *Platoon*, the storyline is focused on the struggle between light and darkness as exemplified by Sgt. Elias and Sgt. Barnes respectively. Elias represents a soul caught in the totally untenable predicament of trying to bring sense to senseless brutality. Barnes has surrendered totally to the darkest position of absolute ruthlessness to keep control. It reaches a climax in a small village which ends with neither concept holding sway and, as such, delivers the final blow to doing what is right. *What is right?* -- that's the phrase pounded into us from our earliest days. It is a way out of the maze of entanglements which, like the land itself, is choking and blinding us.

But where is the light, the purposefulness that will cleanse the soul? What is right? A phrase that never leaves us, haunts us. For many of us an endless, whispering voice.

As the Platoon walks away from the flaming failure of the village, those same strains of Barber's Adagio fill the movie theater. The music creates a literal resonance of sorrow within the heart of the audience. But for an observer who has walked that walk across the chasm of light and dark, the music also fills the mind with the echoes of the past. It reaches a cave of memories skillfully guarded.

As I sit in this moment here in Charlotte, I now am watching a man's innermost struggles play out before me. The music vibrates throughout me. It awakens the memories created by Vietnam and reignited by the movie 'Platoon'. How do you hold the light and not see the dark? How do you pretend it is not there? How can you say that is not in me? That inner conflict played in me then, in Vietnam, and plays in me as well today. I know I am both Elias and Barnes. I also know that I am not alone in those feelings. As I sit in this moment here in Charlotte, I find myself back in Vietnam facing the same dilemma of identity I faced so many years before, and still do so today, and understand Van Gogh's plight completely.

Memories
Alan Brett

Its 11pm and I'm sitting here under the covered part of my patio watching the rain coming down. I look out and see all the trees that surround our property in the back yard. The quiet is interrupted with the sound of the rain, thunder and lightning through the trees. I had to smile as all I could think about was that now I'm dry and no one is shooting at me.

I remember being in Vietnam in the woods with the thunder and lightning and rain so hard it feels like sleet. We are on a mission, and we have been walking since midday. We were to find when a road coming from the north splits off and goes east and west. We were to set up where we had a good view of the split.

We got there as the light of the day was getting dimmer. We found a good observation site where we could see the north, east and west trails. Three men on the east road and three men on the west road. I took my position right above the intersection. The rain was unforgiving, and we had to settle in so we couldn't be seen.

Our mission was to observe who came down the north trail. An intelligence report conveyed that an NVA pay master was going to be on the road. He was to meet up with the NVA cadre in the area and give them the money he was carrying along with orders for an upcoming attack. When we saw them, we needed to call in a head count of the group, what kind of weapons they had, and which direction they took. There were two react teams, one to the west and another to the east to intercept them. We were to follow in case there were any that tried to retreat.

I settled in and got "comfortable" with my back against a tree that also covered my silhouette in case the moon decided to pop out. The rain was relentless, soaking everything I had. What food I had was more like soup and everything tasted like soggy bread. After eating the horrible food, I leaned back on the tree. My gear was soaked. To keep my weapon dry, I covered the muzzle with a rubber because it always keeps the muzzle dry. Also, if I had to shoot, it was like the rubber wasn't there. I wondered, *what the hell I am doing here*. I just ignored everything and continued to keep watch. I only moved to check the other six men to see if they were as miserable as I was, and they were. Someone questioned if we volunteered for this or not. Now back at my tree, I found that my position had been compromised by a hoard of red fire ants. They had invaded my ruck and the only towel I had. I got rid of the invaders and moved to another tree and settled in again.

Just as it starts to get light, here they come. We are all on alert and ready to fire if exposed. First, there were six people we assumed were NVA because they were dressed in civilian clothes. Their rifles at the ready, they walked slowly looking all around. Shortly thereafter was another group of five men, two in front and two in back and one in the middle. I guessed the man in the middle was the pay master. Behind them were four people carrying poles with bundles on them. And behind them were eight more men with their rifles at the ready.

They all stopped at the intersection and moved on to the east. We called in the information, got ready in our soggy clothes, picked up our gear, drained the water out of our rucks and followed them. The NVA got captured and many of the soldiers were killed. All we had to do is wait for a chopper to pick us up.

Back in my yard at home, all I thought of was how wet and miserable I was that night in Vietnam and now I'm dry and no one is shooting at me.

The Mission
Dean A. Little

Earl, silently, begins to pray,
he, his brothers on their way.
In dragonflies, painted OD green,
young men, strong and lean,
skimming fast, above jungle top,
following contour, to the drop.
The flight is new, farther still,
men jumping, to swampy swill,
leeches seek, warm-blooded prey,
soldiers wade, make their way.
Two steps, then hellfire starts,
cutting through armor, body parts,
twenty-three, dropped that day,
none unwounded, in few seconds fray.
Drag each other, to solid shore,
gouts of blood, trail human gore.
Earl crawls up, on pungent land,
bleeding from shrapnel, not to stand.
Brother Fred White, drawing flies,
small red dot, between his eyes.
Earl silently, begins to pray,
He, his brothers, survive the day,
this mission lost, as it began,
men screaming to God and man.
Radio calls bring the rain,
deliver the artillery, gunship's pain,
drive the enemy, to retreat,
time, C4, blows an LZ feat.
Dragonflies drop down, retrieving men,
most-wounded first, lesser then.
Enemy fire, starts up anew,
one chopper left, for the few.

Radio declares, no room for dead,
Can they go up, leaving Fred?
Eyes searching, the others' face,
leave our brother, in this place?
In silent decision, poncho wrap Fred,
chopper waved off, carry the dead.
Earl silently, begins to pray,
as he, his brothers, evade away.
Through miles of jungle, so thick,
walking's a joke, crawling's the trick.
Machetes flail, for holes, no path,
behind, jets, release napalm bath,
drag Fred's body, day to night,
enemy at bay, during flight.
Cloying gnats, fill eye and ear,
Sweat drips from heat and fear.
Insects click, birds caw, monkey's screech,
struggle on, more jungle to breach.
Arms grow heavy, movements slow,
never letting go, body in tow.
Wounds leave, trails of blood,
on branches, vines, in the mud.
Night falls, blind dark drops,
go to ground, talking stops,
rotate vigil, the exhausted sleep,
battle dressings, in silence, weep.
Morning breaks, not their will,
water shared, jungle falls still,
no one moves, then bird trill,
back to, the journey long,
heat, thirst, will is strong.
Another night, next to Fred,
tender care, of the dead.
Earl drifts, dreams of home,
friendly fields, woods to roam,
old dog, walking at side,

stargazing, no need to hide,
Mom, Dad, in loving arms,
cigar treasure box of charms,
the barn, its oaken beams,
wind blows, 'tween the seams,
Dad's dusty, old 64 Ford,
blocked, tarped, a waiting reward,
oh, those hours, to renew,
sanding, painting, make it blue,
remove, replace, nonworking parts,
rewire, gas, see if it starts,
he reaches, turn the key,
hand shaking him, from reverie,
Mom, Dad, old dog gone,
home slipping, to jungle dawn.
On again, grueling effort spent,
nighttime vigils, more sleep torment.
Daylight arises, the sunlight burns,
water's scarce, take carrying turns.
Earl silently, begins to pray,
carry the dead, another day?
Point-man signals, clearing ahead,
chance to end, endless dread.
Throwing smoke, after radio squawk,
no enemy fire or need to walk.
Lifting now, Fred on the floor,
miserable jungle, sliding past door.
The men cheered, welcomed back,
five white men, one dead, black.
Back in the world, not to find,
this ease of being color-blind.
Colonel gives them thanks today,
none behind what we say.
Three days off, then back in play,
Earl, silently, begins to pray.

That Guy: Squirrel Murphy USN - BUSC/SCW
R. Kevin Wierman

To say that Senior Chief Murphy was a character would be a huge understatement. In the Western Iraq Province of Al Anbar that earned the nickname of "The Wild, Wild West," he was brought in as a security and troop safety subject matter expert in early 2005. It was April and I was stationed there with a few dozen Navy Seabees in support of the Marine 2nd Light Armored Reconnaissance Battalion. We were a Detachment from the Naval Mobile Construction Battalion #24 and Senior Chief Murphy was temporarily assigned to "babysit" us to ensure that we all came home in one piece.

Now, I must confess that I had my doubts as I figured that any adult man that had his name legally changed to "Squirrel" could not be playing with a full deck, if you know what I mean. I had only spent a couple of hours with him before I realized that he was in fact in possession of all 52 cards, but it was evident that they had been shuffled, not just once, but numerous times. There was something about his crazy that my crazy immediately related to and I am not ashamed to admit that, for the first several days, I followed him around like a puppy. As we trudged through the camp, we watched the brown baby powder-like substance explode up from beneath our feet with each step. He inspected the berms, kinda like manmade sand dunes, topped with three strands of pulled concertina. The perimeter "needs expanding," he bristled. We scaled an artificial earthen mound and the wooden tower constructed to house one of several machine gun nests. More towers, more guns, more wire and kill zones, one in each direction, and Hesco's, more Hesco's, lots and lots of Hesco's. I tried to write as fast as he could list the methods and details of how we could best ensure personnel safety. Safety was not paramount with Squirrel for that would imply that something else mattered. I was trying to absorb every bit of wisdom and experience that I possibly could.

You see, this eccentric "old timer" had not only been around the block, but around the world. He came to us in the twilight of his career with 2 tours in Vietnam, where he earned a Silver Star, and had numerous security assignments around the globe, but primarily in the Middle East. Most of the guys had no patience for his rambling stories and wrote him off with little or no regard for the gold nuggets that he was leaving at our feet. I would listen to the very end of every tale knowing full well that having been where he had been, having done what he had done and having seen what he had seen, he did not get to be an "old timer" by being stupid or lucky. Squirrel had been known to embellish a bit, take advantage of literary license and add some color to an otherwise routine tale, but I knew that this man knew his shit. All that I had to do was to sift through it to get to the jewels. Most of the other guys mocked and ridiculed him, some while he was even in our presence. They mocked him for his idiosyncrasies, quirks, jitters, flashbacks and numerous other peculiarities. Heck, like Rudolph, he was not even included in most of the "reindeer games." There is no doubt in my military mind that he saved lives, and he did *exactly* what he was assigned to do…*bring each and every one of us home, alive and in one piece.* Senior Chief Murphy died a few years after we returned from Iraq at the age of 65. He finally succumbed to stage 4 cancer, which I honestly believe was the result of frequent exposure to Agent Orange during his time in Vietnam. As for my time in Iraq, he was and will remain, "That Guy" — the guy that made a difference! Regardless of what others thought of him or how they mistreated him, as far as I could tell, he paid them no heed. He just did his job with honor, courage & commitment!

The Worst and Best Day of My Life
Steve "Buck" Owens

On October 20, 1987, I was a 23-year-old Army Chief Warrant Officer Two (CW2) flying the AH-1F Fully Modernized (FM) Cobra Attack Helicopter for the B Company "Bandits" of the 13th Attack Helicopter Battalion of the 3rd Infantry Division located at Giebelstadt Army Airfield. This day was one of the many days we stayed in the field in tents. At this point I had been in the unit almost three years and had seen many such trips. This day was much the same as many other days in that we were conducting "battle drills" against a notional Soviet threat.

The remainder of that day we were following a lead OH-58C Kiowa helicopter that my Commander was flying in, along with another Cobra in front of me. The route we used saw us come within one half mile of the East German border. That October day was very pretty, and we were flying in what we called combat cruise, which allowed the aircraft behind the lead aircraft to sweep left and right as we flew as close to the earth's surface as terrain and ambient light allow, or what is known as "Nap of the Earth." For us, this was about twenty feet or less. I had almost hit a whole family on bikes previously, so this was always extremely low and fast. We flew this way for a bit and decided to climb up to 1,000 feet above the ground for a better view.

We had not been up high for very long when our APR-39 (Radar Receiver) began picking up an air defense radar out our left side, which was not all that unusual. I was in the back seat and had a visual indication as well as an audio tone. I checked to make sure that my ALQ-136 Radar Jammer was on, as well as my ALQ-144 IR Jammer, which are both designed to keep a missile from taking you out. We then began picking up another two sites at the 10 o'clock and 12 o'clock position. Three active sites were unusual, and it got all our attention.

Within a few minutes of getting the radar warnings while flying at 1,000 feet above the ground and 120 knots (140 mph), I heard a very loud bang followed by the aircraft tail moving to the right and the nose dropping to the point it felt like we were going to do a somersault mid-flight. I caught something out of the corner of my eye as it flew by my right door window toward the front of the aircraft. My first thought was that we had been shot down and this would be the start of World War III. I then quickly got back to the fact that I needed to get on the ground quickly. I tried to pull back on the cyclic in my right hand, which controls forward, backward and side to side movement, but it felt like it hit a stop. I pushed on the pedals and remembered they felt like they were no longer attached to anything. I then lowered the collective, which was the control in my left hand that increased and decreased the pitch in the main rotor blades. We began descending toward an open plowed field near the town of Hendungen, which was Checkpoint 21B on the border trace route. I called "Mayday" and told them I was going down. Until they saw my aircraft and realized that the tail rotor was gone, they thought I was making a bad joke. I remember thinking that I was probably going to die, but I was not sad and did not have anything flash before my eyes. I was busy and I still had work to do.

As we descended, I realized that I was not able to control the aircraft very well, and that I might hit a tree line. I called on the radio and stated that I may go in the trees and remember hearing someone say, "stay with it," which was hugely inspirational. As I pulled back as hard as I could to try to avoid the tree line, I remember hearing my co-pilot say to "chop the throttle" which would remove torque from the helicopter. The main reason a helicopter has a tail rotor is to counteract the torque produced by the engine and main rotors. I remember rolling the throttle to idle and the aircraft shuddered and began falling an unknown distance to the ground. I remember just before we hit the ground, I pulled up on the collective all I could and closed my eyes. When I became aware again or regained consciousness, I was hanging upside down from the straps and could see my co-pilot in the front seat hanging limply and I remember thinking that he looked dead. For some reason, the only thought that came to my mind at this point was *FIRE!* I disconnected my seat belt and fell like a sack of potatoes to the roof of the helicopter, and thankfully, I had left my helmet on. When I sat up, I could only see out the one window which had dirt outside of it, and the other window had sky visible.

This next part requires some further explanation. The Cobra is set up like a fighter plane seating configuration of one in front of the other, with the most instruments and controls in the back seat. The gunner's position is where the co-pilot would sit and he had a smaller set of controls to fly, and he could operate the weapons systems. The pilot in the back used a hinged door on the right side of the helicopter and the gunner/co-pilot used a hinged door on the left side. The Cobra was equipped with a canopy ejection system which would blow open the door handles and a linear shape charge would burn out the windows which did not open on the pilots left and the gunners right. This system had a yellow and black striped handle which was located just above and to the front of the collective on the side of the back seat pilot's instrument panel. There was always much fear that it would be accidentally activated in flight, which would be catastrophic. My unit decided to fly with the safety pin in and pull it after we crashed. I only wanted out of this helicopter, so I immediately and instinctively reached up to that little yellow and black striped handle and turned it. It snapped off in my hand as the safety pin was still in. I was shocked, panicked and could not believe how stupid I had been. My immediate problem was that I still had no way to get out of the helicopter. I saw the window on my left that I could see the sky, so I began kicking it. It only took a few bounced blows to determine that was not going to work. I reached for my pilot's survival knife, located on my survival vest and began striking the window as hard as I could with the blade. On the third strike the window broke and my hand went through, deeply cutting my right wrist. I dove out through this hole and stood up moving toward my co-pilot's door. I was able to open it, unfasten him and pull him away from the helicopter. As I pulled him out, he came to and complained that his back hurt. As I stood up. I noticed a noise I had not heard to this point, which was the engine still running at idle. I went back to my cockpit and reached through the hole and immediately rolled the throttle the wrong way as it was upside down. I finally got the engine shut down and I laid on the ground until the Medevac Blackhawk helicopter arrived. They loaded my co-pilot and I into the back and we headed to the hospital. My co-pilot and I held each other's hand for the flight to Wurzburg Army Hospital.

I knew we hit the ground extremely hard and was genuinely concerned I had internal injuries, and I still believed I might die. My co-pilot continued to complain of back pain. I was examined, my wrist sutured and was released the same day, which amazes me to this day. My copilot had a compressed spine, and he was held in the hospital. Upon waking up the next day and getting out of bed I realized that I could barely walk, and every inch of my body was sore and tight. I told my wife that she needed to drive me to work, and she looked at me like I was truly insane. I cannot explain it, but I felt I had to go into work. As I walked through the hangar I saw my Battalion Commander. He came up to me and I will never forget his kindness when he put his arm around me and asked me why the heck I was at work. He told me that we had done an excellent job, and they were putting us in for awards. He filled in the blanks for me and explained that we had not only lost a tail rotor blade, but also the seventy pound gearbox and remaining blade. He also told me not to worry about the aircraft, and said, "they will make more," which I thought was great. Then he got a puzzled look on his face and said there was something he needed to ask me. He asked, "how come you cut a hole in the window of your door?" I thought for a minute and said, "Sir, I'm not sure what you mean." He reiterated that when they got to the crash site, they saw the hole I had cut in the window, but then reached down and turned the handle and opened the door right up. It turns out, I had not cut a hole in the immovable window, which is normally on my left while flying, rather I cut one in the window of the door, which I could have easily opened utilizing the handle. Because it was upside down, they were in the wrong places. He laughed when I told him, and I nervously laughed thinking about the crap I would take not only from the accident investigators, but also my fellow unit members.

The accident investigation interviews occurred within a couple of days of the crash. When you think of this board, think of the most intimidating job interview in front of six to eight people, including a

Psychologist, Instructor Pilot, Flight Surgeon, and others. The board concluded a few days later and the results came some months later: metal fatigue caused by cracks starting from the inside on the tail rotor yoke, which is what the blades attach to. Turns out this part had been in a previous accident, and it was cleared for use again. We were awarded an Army Commendation Medal along with the "Broken Wing Award" which is given by the U.S. Army Safety Center for actions during an inflight emergency. The moral to this story for me is that I should have been dead at twenty-three. Oh, and yes, I received lots of good-natured crap over cutting the hole in the pilots' door window.

This was the best and worst day of my life for a couple of reasons. I think it made me a better pilot overall, but it also made me appreciate how precious life is and how unforgiving aviation can be. Sometimes when I am feeling low, I remind myself of how many "extra" years we both got. There I was on October 20, 1987, and here I am now talking to a VA counselor about the feelings that this day as a young man still brings up 37 years later.

When You Really Have a Team
Donna Culp

Whether mentoring or being mentored, we worked to provide an honest, ongoing assessment of where strengths and weaknesses were, and we strived to do our best and give our best.

We could practically read each other's minds. That was the first time I had ever experienced that level of teamwork. Very rare and very much appreciated!

I was privileged to mentor three enlisted team members who were eligible to sit for the National Occupational Therapy COTA (Certified Occupational Therapist Assist) certification board exam, and all three made it.

Over the years since leaving the military, whether facing challenges or reveling in overcoming those challenges, the yardstick always falls back to these dedicated people.

Terrorists
Michael D. Hebert

Terrorists have been active for decades, but not as far-reaching as in the past 20 years or so. Some of us remember the odd skyjacking, some ideological person or group takes over a plane and demands to be taken to Cuba or elsewhere. Sometimes they kill someone to make a point, but by and large they were seemingly random, isolated incidents.

Organized terror groups started really raising their heads in the 1980's, with Hezbollah, Muslim Brotherhood, Abu Nidal starting to launch random attacks, increasingly effective and deadly, in addition to kidnappings of Westerners, particularly in Lebanon. Not all of them were Islamic groups. There were a few non-Islamic ones, like the Basques in Spain, the FARC in Colombia, etc., but the ones most deadly to Westerners are jihadists preaching Islam in its deadliest form. With time, experience, and funding, they have become the threat we have been facing, prior to but more deadly since 9/11. Al Qaida and ISIS started gaining power in the 1990's, but since 9/11 we have been perpetually fighting them on several fronts, and their activities have substantially changed our lives. They've changed how we board flights, how we train to fight them, and how we collect intelligence on how they organize and fund themselves.

In my experience, terrorists are a tougher lot to fight than the Soviets. Terrorists are happy to become martyrs, so it's a different animal to take on an enemy which would be happy to die in the name of Allah. They're also less inclined to discuss terms of peace, ceasefires, etc., since they're not interested in negotiating peace as long as us infidels are still alive.

It is frustrating to me to see how much of the world is now under jihadist or Islamic extremist control, despite our commitment of military forces, lives, finances and other resources. I've worked with many governments fighting terrorists, and in many cases the terrorists have won out. Examples: Afghanistan, Mali, Burkina Faso, Niger, Lebanon, Yemen. Isis and Al Qaida are very active in southern

Africa, as well as South and Central Asia, and there's no indication they will not prevail there as well. Boco Harem in Nigeria kills Christians by the hundreds, and that's hardly reported in the U.S. media. They are entering the U.S. in large numbers, and I don't believe it is to embrace our way of life, settle down, get a job and pay taxes. I think we need to remain vigilant of extremists in our midst and be wary of any plans for attacks against our citizens and our way of life.

What Can Happen?
Richard Epstein

Do I choke him out or walk away?
I think about that as I sit next to him.
Funny: he has no idea what I'm thinking.

When I tell him I'm on my way
to Southeast Asia, he looks at me and shouts:
"You're going to kill babies!"

He's my cousin.
He's younger.
It's our first meeting.

I'm not used to the idea of killing.
I hope the need won't arise.

A cook doesn't kill.
An electrician doesn't kill.
A carpenter doesn't kill.

There are many jobs in the army
that don't require one to kill.
But I'll be alert and prepared.

Something can always happen.
It's up to Lady Luck.
That's what I think.

First Time Outside the Wire
Tom Baker

I made my choice.

My buddy Bledsoe and I graduated from Parachute Rigger School at Fort Lee, Virginia, in early February of 1967. After a two-week leave, he and I were headed to Vietnam. We arrived in Cam Ranh Bay on the 21st day of February. The 68 Tet had been going on for three weeks.

I remember getting off the plane with the heat and smells hitting me like a brick. But it was the look on the faces of the GI's waiting to board our plane, their Freedom Bird, that shook me the most. The stare, not all, but many of them had a look of having seen and done things that no one should ever have to see or do. Hollow-eyed, gaunt features told me that if this was what I came here for, it was out there—in the heat, in the rice patties, in the jungle, and in the hell of a fire fight. Be very careful of what you wish for, GI.

We took a C-123 from Cam Ranh to Phu Bai with several grunts headed to Camp Eagle, main headquarters of the 101st. We checked in at the orderly room and settled into our hooch, which was an old French Colonial building. Next day, we began our job of rigging out loads for resupply to Khe Sanh, which was getting hammered about a hundred miles northwest of us.

At that time, 2nd Battalion of the 501st was near Phu Bai and getting its ass kicked. Colonel Megers, the Battalion Commander, needed and called for a QRF (Quick Reaction Force). Bledsoe and I raised our hands. Anything to get out of this rear echelon shit. At least that's what we thought. One night, my seventh day in country, a Special Forces Compound, which lay a few kilometers up the valley from our airfield, got hit hard. The QRF was called up. "Get your shit together, people, we are moving out shortly."

It wasn't long before our platoon sergeant stepped into the hootch. "I want you people to write your last will and testament letter, who gets what and all that shit. Leave it on your bunk when you finish." Talk about a sobering moment. One by one we finished that heavy task, walked outside, and sat on the sandbags that surround every hootch.

Exactly ten minutes later, we were headed for the main gate. Now, I will tell you, the first time you step outside the "wire," in Nam, it is a scary feeling. But what is the spirit of the bayonet? "To kill." That is what. We walked along both sides of the road. Sergeant Moore was walking point with the LT and his RTO walking ten yards back. We followed like school children.

We were doing a sweep on a village that had probably been in existence for a hundred years. Early morning light chased the ground fog away. My eyes darted everywhere and saw nothing. Chickens, hogs, a few old people, smoking firepits, and poverty. No VC or they could all be VC; how would we know? *What the hell am I looking for*, my mind screamed as we walked into the village. The M-16 was hot in my sweaty hands. My heart beat a little faster as I saw motion. My weapon moved in the direction of a small hut.

Out of the smoke from the cooking fires, a small child walked out of the hut—a girl, maybe three or four years old. Her face round, her eyes bright with innocence. We all had heard the many tales of how the VC would strap grenades on children and then somehow blow them up around soldiers. Was this little girl a weapon? I stood dead still as the child walked toward me. Sometimes in combat, you have to make decisions about life and death in a heartbeat or less. I made my choice that morning. I was not going to kill a child just to save myself.

Bledsoe was okay, twenty to thirty feet away, so it was just me, the little girl, and whatever god puts people in these situations. I shifted my rifle to port arms across my chest as the child walked up and stopped about two feet away. "Okay, this could be it," I thought. The scene may have been comical to the gods of war but not to me, until she smiled. Hell, you could have knocked me over with a feather. Our eyes met, my heart melted, and the world stood still for a second. I barely heard Sergeant Moore's voice shout, "what the hell are you doing, Baker? Give her a Hershey bar and let's move."

Neither me nor the child looked toward Moore, but I remember taking a Hershey bar out of a pocket, kneeling down, and handing the chocolate bar to this tiny piece of life. I didn't know what to say, so I just winked at her and smiled. She smiled back, then turned and walked back into her hut. I didn't search that hut. Was this whole incident meant to keep me out of that hut? I'll never know. I just hope that child remembers at least one friendly American soldier.

A Life Spent
John Mason

    He enlisted in the Marine Corps on a 4-year hitch. Training for West Pac at Camp Pendleton and after listening to troops just rotated back from combat, he was afraid but volunteered to be sent immediately to Vietnam and the next day shipped out.
    For 13 months he served as a grunt radioman, first for his platoon, then for the Company Commander. He was meritoriously promoted twice, making corporal. Humping the radio, he was prime target for VC/NVA and although not wounded by enemy fire, he suffered a non-battle casualty from anaphylactic shock due to centipede bite, to trench foot, to jungle rot, to malaria, to infections from leech bites, to heat stroke. And when his tour was complete, he extended in country for another 6 months where he was sergeant-of-the-guard for DaNang Air Base and was recruited for Drill Instructor School. He rejected D.I. School because, as he told the C.O., he didn't want to train Marines to do what he did in Vietnam.
    Back in the world he was a loner, drank too much, couldn't adjust, resented stateside discipline and somehow got crosswise with his First Sergeant. A buddy from his Pennsylvania hometown could take him home one weekend but his First Sergeant denied weekend liberty. He went anyway, UA. Back on base he got office hours and was busted. Then on KP duty with green troops, "boots" who never served in Nam, never did what he did in combat for 13 months, never extended, he went UA again. Two months later he returned to base, this time in the brig, charged with "desertion" and offered a deal. Get back to being the twice meritoriously promoted Marine, don't go UA again, serve the rest of his enlistment, and the charges would be dropped. Would he swear not to go UA again? He answered honestly, "No sir," he said, he just didn't want to be a Marine. He couldn't take it anymore. Then, would he accept discharge "for the convenience of the service" and waive VA benefits in lieu of court-martial and prison? He did.
    As a civilian he couldn't hold a job, was addicted to alcohol, then

drugs, had failed marriages, nightmares, isolation, cancer, another cancer, kidney failure, dialysis, depression, desperation and was anti-social and homeless.

Trying to clean up his life in rehab, he called his squad leader from Vietnam. Could he help with a discharge upgrade? Yes, and twenty-five Marines, including his Company Commander wrote to the Bureau of Personnel reporting his courageous service in combat. The upgrade was rejected. Later then, living on SSA benefits, diagnosed with PTSD and with both cancers, PTSD and kidney disease being presumptive illnesses arising from Vietnam combat stress and Agent Orange poisoning, he tried again for an upgrade which would allow VA benefits, and while the matter was pending, he died. His life spent by Vietnam.

What I Learned from the Vietnam War
Larry Kipp

I was a nobody who didn't believe in myself when I first went to Vietnam. When I graduated from Medic School, I was a theoretical medic. In Vietnam I learned to become a real medic. It was only after saving my first life I realized that even if I do die the world will continue to be affected by something I did. I came to believe in myself. I extended my tour in Vietnam by another year because saving lives seemed like a worthwhile thing to do.

I saw a lot of mutilated people as a medic over there, some beyond recognition, and some were only in parts. These events have given me some pause when I consider my own problems, and at some point, I realized that "problems" is not a category, but a designation. So, problems could be categorized, and I came to understand there were at least two categories of "problems": GOOD PROBLEMS and BAD PROBLEMS.

I knew what the "bad problems" were: broken backs, split skulls, dead children, and crispy burnt people, to name a few. This gave me, then, some perspective about what I might call "good problems": foggy glasses, a speeding ticket, missing lunch, a flat tire on my way to a job interview, and a burned-down apartment, to name some. Then it dawned on me that when I am first hit with a problem, before I start hyperventilating about it, I should categorize the problem by asking "is this a good problem or a bad problem?"

I became amazed at how asking this question, at the start, really let me be at ease because nearly ALL my problems could be classified as "good" ones. I also discovered that "being at ease" when attempting to address a problem often led to better results. Looking back, I have to say that all but a few of my problems have been "good," since they were all far better than being split open or losing a child.

In war, when you have a problem, the last thing you want to do is panic. Panicking doesn't make the problem go away and doesn't solve it either. Panicking usually made the problems worse. I learned, when faced with an unsavory situation, the only thing you can do is WORK THE PROBLEM. That's what I chose to do, and it worked every. single. time.

We each have our own set of problems, and I admire those who have faced what I might call a bad problem and have continued with their life making do and adapting. In each case, I become filled with awe at just how adaptable these folks are. I can only hope I would be as adaptable were I in their shoes.

Making choices: When you are approaching a hot LZ (where you are being shot at), you have to weigh the risk of completing the mission versus getting shot down. Completing the mission is key. But if you get shot down you 1) don't complete the mission and 2) will lose the chopper, which means no future missions will be completed either. Plus, there is always the possibility when getting shot down that we may not wake up tomorrow either, but that really is secondary. So here you are, you've got wounded on the ground, some of whom may die if not taken to a surgical hospital ASAP versus the possibility of not completing this or any future missions. There is never enough information available to make a precise decision, so you have to go with what you do have. First, we can circle the LZ to get a lay of the land and make an estimate as to the best approach and departure, or we can get that information, or at least part of it, from the radio operator on the ground (but they may not see, or know of, certain land features that the Aircraft Commander (AC) can see and exploit). In a few dicey instances, the AC asked us in the back for our opinion, we always said we were ready when he was. One nice thing about the Bell Huey is that it can take a lot of hits without harming any key systems, but one shot in the wrong place can bring one down in a hurry. I never flew on a mission where our AC declined to go in. And so, in we would go. In all but a few instances we completed our mission: we picked up the wounded, patched them up, and got them to a surgeon. In a few instances our chopper got so damaged that we could not land and had to abort the mission. Explosions, trees, close enemy fire, and loss of night vision can abort a mission in a hurry. Those really sucked.

In the long run, deciding to go in is, in every case, based on unique circumstances that cannot all be predetermined. Pilots only have their training, their personal experiences and knowledge, and their gut to go on. It's a hell of a way to fly, and not all of us came back. But the first goal is, always, "complete the mission."

Our F Troop, 8th Cav "Angel"
John T. Hoffman

In early October of 1972, as the Blues Platoon Leader and helicopter pilot, I led a heavy team from F Troop, 8th Cav from DaNang to the old airfield down in Chu Lai, well south of DaNang Air Force Base. Our mission was to interdict North Vietnamese Army (NVA) forces trying to infiltrate into the area in southern Quang Ngai Province from the mountains to the west. The NVA had been moving small units into the central portion of the province so that they could move against the US Air Force base just outside of the city of DaNang. Our mission was to find and engage these enemy forces in order to minimize their ability to attack the air base with indirect fire from rockets and mortars, as well as direct ground attacks on its perimeter.

We moved into the former home base of the troop, before its move to Marble Mountain Army Airfield in 1971. At this point in the war, the Chu Lai Airfield was now an Army of South Vietnam (ARVN) base with a US special operations unit occupying a small part of the base that was secured by a combination of US and ARVN security forces. Our F/8 Cav team was comprised of two UH-1H Hueys (Slicks), three AH-1G Cobra Gunships (Guns) and a small maintenance team. Also with us was a reinforced Blues squad from our Blues platoon for our own security and to assist in any downed aircraft or crew recovery operations, if needed. The "Blues" are the organic infantry within each air cavalry troop.

We located ourselves on the old Rosemary's Point compound, high on a bluff over the South China Sea. It was the most secure location on the old base and the quarters for the former base commanders when this was a major US operating base prior to 1972. We occupied several US style bungalow homes on a cul-de-sac behind some heavy wire security barriers. Each small home had a driveway from the access road to the front of each structure. As a security measure, we did not leave our aircraft down on the airfield. We parked them right in these driveways. So, we slept close to them to facilitate rapid aerial exfiltration, if needed.

We operated both day and night, depending upon the intelligence available on enemy activities or combat events that occurred. Our night operations, referred to as Night Hawk missions, were conducted in the dead of night, with our aircraft blacked out, in order to catch NVA forces out in the open, moving or preparing to conduct infantry, rocket or mortar attacks. Our night vision system in 1972 was the Mark-One Eyeball. Operating blacked out also reduced our visual signature and made targeting our aircraft much more difficult when we flew just above the ground or treetops. These nighttime missions were the rule during this remote deployment from the rest of the Troop at DaNang Air Force Base.

One night we learned of a suspected impending attack on an ARVN security position south of the air force base. We launched at midnight with two guns and my Slick. I was the Air Mission Commander for the mission. In addition to commanding the small force, I coordinated our operations with supporting US Air Force assets and the ARVN forces on the ground. There was also a US Advisor with the ARVNs on the ground. We arrived in the vicinity of the ARVN position to observe sporadic small arms fires being exchanged between the ARVNs and the NVA force on the ground. As we flew at near treetop level, due to the threat of shoulder fired Soviet SA-7 anti-aircraft missiles, the enemy small arms fires often passed around, below, behind, even sometimes above us. Initially, we had some difficulty locating the main force of the NVA on the ground, as they seemed widely dispersed based upon the small arms fire we could observe, whose tracers were most often green in color. US supplied ammunition employed red tracers, so we could normally identify the enemy force by the color of the tracers, especially at night. But not always. The enemy also employed captured US weapons and ammunition, so sometimes we observed a mix of red and green tracers from suspected enemy positions. The ARVNs were careful to only fire US ammunition when we were supporting them, so as to prevent friendly-on-friendly engagements.

This dark night, the green tracers were flying through the air with unusual intensity. The NVA could hear us but could not track us in the dark unless we flew right over them. When we did, we flew at high speed to reduce the time window the NVA had to engage us. So, most of the NVA fire intended to hit us actually went wildly wide of us and filled the sky around our aircraft with green streaks of light, something like a 1950's science fiction movie. It was very unnerving and all very dangerous. But we focused on the mission at hand: find the main enemy force and engage it so that the ARVN forces could accurately target it with indirect fire. The bad guys knew our gambit well and exercised excellent dispersed fire discipline in order to not give away their main force location. So, we flew over the suspected locations and began to engage positions we could identify based on enemy AA fire and observed forces on the ground.

It was not long before we found a large force of NVA that were moving toward the ARVN position. We circled back over the location and my door gunners, on both sides of the aircraft sitting in the open doors, began to call out fires and enemy elements on the ground. I was carrying extra crew in the back of our aircraft. The more eyeballs observing the better and it enabled additional weapons to engage the enemy, if needed. In this case, I had two manned, door mounted machine guns, one an M-2 50 caliber heavy machinegun and the other an M-60s. I also had one extra gunner, Jim Fentress, sitting in the left-hand cargo door, directly behind my cockpit seat, with an M-60 machine gun on a shoulder sling and an M-79, 40mm grenade launcher close at hand. The M-79 was useful for engaging enemy anti-aircraft positions and hardened targets below us.

As we flew back over the enemy force the crew called out enemy firing at us on our right and I banked in that direction to observe the location. This would enable me to call out that location to the guns with us. Just as I did that, the crew called out heavy fire coming at us from the left. I quickly and abruptly banked back to the left in a steep turn to allow the two '60s on that side of the aircraft to return fire and to aid in directing the guns with us to engage the enemy firing at us. As I got well into the steep turn, I heard one of the crew yell out over the intercom: "Holy shit… Fentress just fell out!" Simultaneously, I felt the sharp tug on the controls as the aircraft's center of gravity shifted to the left. I was well into the left turn, as the monkey strap attached to Fentress went taunt at the bottom of his fall. I looked back and, sure enough, Fentress was gone from the cargo compartment doorway. "Oh crap" I thought, he must be hanging under the aircraft about 4-5 feet below us. I flattened the turn and yelled over the intercom to pull him up. Then I heard firing of the M-60 firing below us and saw the flashes from the muzzle blast of Fentress's '60 down through the aircraft chin bubble below my feet. I began to lift the aircraft almost straight up as I pulled in collective with my left hand to increase lift in the rotor blades to gain altitude and prevent dragging Fentress through anything on the ground. This would also help reduce the ability of the enemy to see and target Fentress himself. In fact, there was no ground fire that I saw coming up at us at that point as we rapidly rose up into the dark night sky. Almost immediately, Fentress was pulled back into the aircraft by the other gunners in the back. He plugged back into the intercom and excitedly said "I fell out in the steep turn……but I am okay. Sir, let's not do that again." Thank God for the monkey strap that saved him.

    I was truly relieved and silently thanked Jesus for saving him. But I also had to refocus on the mission and get the guns engaged on this enemy position. Then, almost in unison, every NVA soldier below began to fire at us. Tracers went everywhere and it was time to pull back. The guns expended their rockets and "chunker" fire from their chin mounted 40mm gun on the general area of the enemy force, as we pulled up slightly and away from the intense ground fires. The enemy force gave away their main position and friendly artillery fires were now inbound, called in by the ARVN observers in nearby ground positions.

    It was about to be a very bad night for the NVA attack force below us. It had nearly been so for us as well. Jim Fentress was back on board, and we had no casualties. We did have, we found later, a few new holes in our Slick. We were all quite unnerved by the night's turn of events, despite our overall mission success. Jim Fentress had a new nickname after that night: "Angel." We still call him that. I often think about what it must have been like for him to fall into the midst of the enemy force, only a few feet above the ground, at night, in the middle of a battle. What a ride that must have been, even if it only lasted a few, very long minutes. And what did the

NVA troops on the ground think when a door gunner dropped out of the helicopter into their midst firing an M-60 machine gun as he flew though the sky just above them like superman! It certainly must have stunned them, as they momentarily did not engage him or us. Angel's fame spread fast through the troop and was even written about by an AP reporter, Richard Blystone, who visited us while we were operating out of Chu Lai for that mission.

To read the rest of my story, look for my book, *The Saigon Guns*, at your favorite online book seller.

Ode to Donald
Carl T. Zipperer

We met on the flight line 'cause you crewed 239
We flew together you and I and we did just fine
Old 239 went home now 759 was your baby
We carried heavier loads and we never said maybe

Can we get in there? -- It looks so tight
I just waited to hear clear down left, clear down on the right

Our jobs weren't easy -- nor were they fun
Although they would have been if we weren't catching rounds from a gun

We were young but always ready
For you and I had to stay steady
We might take fire at any place
Now, tears still trickle down my face

We became good friends in a very short time
Where days seemed like years and years became a lifetime
We spent hours together, just you and I
We became friends on the ground and a combat crew in the sky

You came from Germany with just six months to go
Who would do that to you, just why -- did you know?
He would teach you a lesson, try to make you a man
So, he ordered you off and away to Vietnam

We were young but always ready
For you and I had to stay steady
We might take fire at any place
Now, tears still trickle down my face

Then, Don, you flew where I didn't go
It was Lam Son 719, I was recovering, you know
On missions that seemed hopeless in thick smoke and haze
It was late March during those final crazy days

Our losses were great up above Route 9
108 choppers lost in such a short time

72 aviators were lost when the enemy hugged friendly bases
Our hearts were so heavy for 72 ever-young faces
Of the crewmembers who died with 11 men not found
Your ship's crew of four never homeward bound

We were young but always ready
For you and I had to stay steady
You fell down on fire at that far-off place
Now, tears still trickle down my face

Tent City
Sarah Scully

Rows of giant Army tents stretch over a gravel lot laid out in a desert landscape. In between each tent, towering lights loom overhead, grounded by air conditioning units pumping into every tent and powered by roaring generators operating 24 hours a day.

Every tent city looks the same. Feels the same. Overpowering, existing with a bleak and regimented precision. Plopped down wherever the military needs Soldiers, the tent cities house the Army for the fight.

Female and male tents group together neglectfully in the center so female Soldiers must carefully navigate their way through the tent city to their bathrooms and showers along the perimeter.

It's almost midnight.

The other females are asleep in bunk beds lining the tent interior. I hover anxiously by the tent opening, peering out into the darkness, debating within myself.

Should I wake someone up, bringing a buddy with me…or risk it?

I hear the crunch of boots on gravel nearby, and freeze, like a startled rabbit in the brush. After a moment, I go over to the closest female and gently touch her shoulder as her eyes pop open.

"I'm sorry, so sorry," I say in a quick whisper. "I've really got to go, and I need a buddy. Nobody else from my unit is here."

She's instantly alert, eyes free from previous sleepy haze. She sits up quickly and slides on her shower shoes.

"No problem. We've all been there. Not safe otherwise," she whispers, as we walk out of the tent. She casts a hateful glance at the generators and a wary glance at the male tents.

"Out here, no one can hear you scream."

We make it back to the female tent and fitfully drift back to sleep. I feel a tentative touch on my shoulder and startle, instantly awake.

It's a young female.

"I'm sorry," she says in a hesitant whisper. "Can you go with me to the bathroom?"

Home
Wallace Bohanan

We were two-and-a-half months into a one-month rotation on the DMZ. The hot, dusty Firebase was taking incoming mortar, artillery and rocket fire daily. I had stopped thinking of home, despite the occasional care package my folks sent me. The reminder that there was a safe, orderly place that I could possibly inhabit made the hell hole I was in even worse.

I longed to be able to stroll down to the Sugar Bowl, purchase a milkshake, sit at the counter, and leisurely sip heaven through a straw. My friends would be there, and we would laugh, tell jokes, and flirt with the fine young ladies in miniskirts. The air conditioning would be blasting while we planned where we were going to drink, smoke pot, and dance the weekend away. But I had to stay alive in order to do that. Best not to think about home when we were putting rounds downrange during a fire mission or scrambling for shelter when incoming rounds were whistling our names.

The welcome home did not meet my expectations. Not one cab would take me home from the airport. The three-hour wait for the early morning bus to the subway was excruciating. My dress greens and duffle bag elicited no response from the passengers on the "A" train as it roared towards Brooklyn. I was bone tired, yet excited, as I walked the eight blocks through the concrete jungle that led to the projects where I lived. I could barely contain my excitement as I rode the elevator to the eighth floor and knocked on the door of apartment 8A. No answer. I knocked repeatedly to no avail.

It was 8AM, and the projects were coming alive. The sun was shining, and the early rising neighbors were out and about. It was exhilarating to stand in front of the building and talk with folks I knew while growing up. No one asked me "how was Vietnam?" They were just glad to see me, and I was joyously pleased to see them.

It was 10AM before my younger brother finally awakened and answered my persistent knocking. When I walked through the door, he gave me a big hug and handed me a joint. My soul soared as I gratefully realized I had made it home.

The Cold War
Michael D. Hebert

After Vietnam, the challenge for the U.S. Armed Forces was still the Cold War, training and preparing for the possibility of an armed conflict with the Soviet or other communist forces in any environment. In my case, not only did we plan for conventional and unconventional confrontations with Soviet forces, in Naval Special Warfare we trained with tactical nuclear weapons we could deploy, as well as a range of tactics in a wide range environments, including deserts, mountains, jungles, urban and arctic.

My favorite specialty in the Navy SEALs was under-ice demolition. The rationale was that Russians/Soviets preferred to drive their convoys over thick ice than through forests in a winter environment. We learned to place demolition charges under the ice – to blow as an enemy convoy drove across, sinking the convoy into the freezing water. A very efficient way to take enemy combatants, as well as their equipment. It only takes a few small charges to disrupt the integrity of the ice. Similarly, a few small charges can cause an avalanche in snow above a road traversed by the enemy.

Another factor in winter warfare is that the supply chain is much more crucial for the survival of soldiers than in a more temperate climate. Attacking enemy vehicles, by stealth using snowshoes or snowmobiles, can have a strong effect against a larger enemy force down the road. The troops can freeze to death if not supplied with food, water, or transport.

We spent a tremendous amount of effort, funds, and lives trying to thwart the spread of communism. We built a massive arsenal of weapons, including nuclear. We devoted a large number of resources to monitoring the activities of the Soviets. Their submarines and bombers patrolled our coasts constantly, and we kept close eyes on them. Our troops in places like Europe, all along the Iron Curtain, were in danger of close encounters that became deadly from time to time. They had to be vigilant all the time.

I felt bad for the people who were subjugated by the communists.

We saw a lot of relief when the Soviet Union broke up, and smaller countries were free for the first time. I was in Germany when the Berlin Wall came down. Thousands and thousands of East Germans and folks from other liberated countries in the Eastern bloc were driving down the German autobahns, yelling and waving their arms out the windows of their little Trabants, the small puny cars made in East Germany out of wood. I had the great opportunity to travel to several of the liberated former Soviet countries, providing training courses to their new governments, and seeing how happy they were to be free. Many of those countries have since joined the EU and NATO. Ukraine and Georgia are two of the countries Russia is trying to reclaim by force, and those citizens want no part of it. Moldova has an enclave that was never relinquished by the Russians. There are enclaves in the Baltic states – Latvia, Lithuania, Estonia, still controlled by the Russians against the will of the local government and people.

Other countries with substantial Soviet presence suffered as well, even if they were not part of the Soviet Union. Ethiopia, for example, managed to kick the Soviets out after suffering from their presence. Before they left, took every piece of machinery they could and shipped it back to the Soviet Union. They even took all the animals from the zoo in Addis Ababa. They basically looted the entire country before leaving. Afghanistan suffered when the Soviet Army invaded, killed their leader, and raged a brutal war before having to withdraw in defeat. The Afghans had a strong economy and peaceful existence before that invasion and have never recovered since. Cuba and Venezuela are still under the influence of the Communists. Their economies are suffering, and they are hostile to us, the United States. Not healthy in our backyard. We must never forget how damaging and expensive the Cold War was for us, and we must now be wary of a Cold War II.

Eyes that Talk
Richard Epstein

He had a puppy that followed him everywhere.
One would guess he lives in the village outside the wire.

Always smiling, always playing with that yellow pup.
Every day he comes through the tangle wire.

We made a hiding place for him under a bunk.
If the Lt found out, he'd be a goner.

In the meantime, we feed him. We play marbles
in the dirt. He learned to spit shine our boots.

He learned to play cards. We read him stories.
He doesn't understand. He doesn't say much.

But those eyes; they talk.

Arrival at an Anti-Communist Guerilla Camp
Tom Hickey

    We turned off the highway onto an unmarked side road. After a while we turned off the side road onto a narrow dirt road. The terrain became hilly and heavily wooded. After another twenty minutes as we turned a curve, the driver slammed on the brakes. Ahead on a hill by the road was a cattle truck with heavily armed soldiers. "Get your gun!" the driver yelled. "Get out your gun!" Although we were smuggling five M1 carbines under a tarpaulin in the back of the jeep, I was the only one actually armed at the moment. The truckload of soldiers dismounted and headed toward us. I got out my trusty little snub nose .38 to confront the first AK-47 I'd ever seen. Its distinctively curved banana clip and wooden stock were a reddish orange color. I thought that was a strange choice for camouflage in dark green jungles. "It's okay!" the driver grinned in relief. "They're ours!"
    A guy who seemed to be in charge stuck his head in the window and asked if the doctor was on board. The driver pointed at me and the officer said, "Take him straight to the camp. We've got some real sick men."
    It was getting dark. Our dirt road shrank to a narrow trail in the woods. We barely fit as branches battered both sides of the jeep. The hills got steeper and closer. A man stood among the trees, watching us. He wore a green fatigue cap, a woodland camo shirt with the sleeves up, dirty blue jeans, a black eye patch, and cradled a 12-gauge shotgun. Other armed men were scattered in the darkening woods. We approached a small light and a low machine noise. A light bulb strung between two trees hung over six men writhing on the ground. Laying in a row, they wore camo fatigue pants and some had T-shirts. Their groaning fought with the muttering of a tiny Honda generator powering the light bulb. I got out of the jeep with my green metal aid kit. The guys were pouring sweat, hot to the touch, thrashing around and delirious. Most groaned with muscle aches but a couple of them were pretty out of it and didn't make much noise. I figured malaria and dug out syringes, alcohol and a sort of injectable aspirin which

quickly relieves pain and fever. After injecting each of the sick men, I presented myself to the camp commander.

"Diego" was clean shaven with a black beret, green short sleeved fatigues, and a venerable Browning Hi-Power 9mm pistol on his hip. He introduced me to "Jupiter," his XO. Jupiter was an older dude with a moustache, green fatigue cap and trousers, but the T-shirt they had given him was PINK! I would have felt sorry for him except that he had the coolest weapon in the camp: an Israeli Gali assault rifle. By the time we finished introductions and a brief discussion of the mission and makeup of the camp, the sick men had stopped shaking and fallen asleep. I arranged with the CO to get them some chloroquine tablets the next day. They then assigned me to bunk in the officers' tent. Someone walked me up a hill to a black tent in the darkness under the trees. The tent was made of thick sheets of black plastic stapled through cardboard squares to wooden poles cut from local trees. The men had used their knives to cut big square holes for windows and doors. Our rows of "champas" (hammocks) were made by hammering wooden poles into the ground and nailing crosspieces onto the poles at the head and foot ends. Coffee and bean sacks were stretched between the crosspieces to sleep on. I had a blanket but the cold air under the hammock wouldn't let me sleep until I lay my towel under me. It was still a little damp from a quick open-air shower I took before bed, but gave some insulation.

Early the next morning, assembly was called. The CO and XO stood beside a flagpole. The men were lined up in rows facing the officers. I was parked with the men. Some of the men had weapons, but apparently there weren't enough for some, including the two women. Hopefully the five M1 carbines I'd brought in would help. They ran up their country's flag and sang their violent national anthem. Then the CO shouted, "Have the doctor stand in front of us!" All these armed men stared at me grimly as I walked up and was stood in front of the flagpole. *Shit! What did I do?* I thought. *I only just got here and already they're gonna shoot me?* Then the CO shouted, "Let's all applaud, because now we have a doctor!"

Fond Memories
Alan Brett

I can remember several times that I found myself in a place that took me away from being in the middle of fighting a war.

One time, when we were by one of the rivers in the southwest area of Vietnam, we came out of the jungle into a clearing. The first thing I saw was a row of sampan boats lining the beach. Then the village came into sight. As we entered the village, we watched the villagers who were busy either making boats or painting them in bright colors. The entire village was involved in producing sampans to be sold.

This was an amazing sight that took me out of the war. It seemed that the village wasn't part of Vietnam or the war. The colors and what they were painting were so beautiful that I stared in amazement that they were able to do this with all the confusion around them.

Another time, we were in the mountains and came across a village that had rows and rows of kilns with smoke coming out the chimneys. The women and older children were painting pots as fast as the older men and women could get them out of the kilns. Their painting was also in bright colors with many different designs. Again, I was stunned at seeing something I couldn't believe was going on in the middle of all the craziness of war. The pots were so pretty, and the villagers appeared to be happy and peaceful. It took me back to my own attempts in making pottery when I was a teenager.

Every time we were in helicopters flying out for a mission, I couldn't help but notice how beautiful the landscape looked from high up in the air. It was amazing how the trees and/or the sea line looked so beautiful. And how the terrain changed as we went from rice paddies to jungle or from shore to the blue of the water. It took me back to being a lifeguard or camping with the scouts in the states.

These visions didn't last long but were a great relief from what my day-to-day life was then. I still remember them with fondness.

Boom
Harold (Ted) Minnick

The definition of a story, according to Wikipedia, is: *"a narrative, an account of imaginary or real people and events; a piece of prose fiction that typically can be read in one sitting."* We all remember as children when somebody would "tell a story," it usually meant they were fibbing or lying and usually they would end up with a paddling or some other punishment. Another little tidbit is how you can tell the difference between a fairy tale and a war story: a fairy tale starts out "once upon a time" and a war story starts out "you ain't gonna believe this."

My story today is called *"BOOM"*.

Once upon a time, there was an Army field artillery composite battery comprised of 8" howitzers, 175mm guns and various other track vehicles. Alpha Battery, 6th Bn, 32d FA was in Darlac Province of II Corps, very near the Cambodian Border and the Ho Chi Minh trail. We were approximately 120 miles west of our higher headquarters at Phu Hiep, just south of Tuy Hoa Air Base. Since it was far from HQs an OH-6 (Loach) helicopter (looked like a teardrop resting on its side) was attached to us primarily for aerial observation of fire missions, route recons for convoys and other necessary missions. The aviator assigned to this aircraft was a 20-year-old Hispanic Warrant Officer who spoke with a heavy accent and went by the name of Andy. Andy flew this aircraft like it was attached to him. Andy kept the aircraft at the DarLac Province Headquarters because they had revetments, good hooches, great chow (not C-Rations like we had) and a small PX for Class VI items such as liquor and cigarettes. Whenever we needed to fly a route recon or shoot a fire mission into the thick triple-canopy jungle we would use Andy.

Andy's philosophy about flying around our area of operations was to fly treetop level and very fast. Didn't make any difference if we were over rice-paddies, the jungle, or the South China Sea. He said that way by the time Charlie saw you, you were too far away to shoot at. Made sense to me even though I held my breath anytime he was hauling me somewhere.

Andy always traveled with an M-79 grenade launcher (Thumper or Blooper) and 5-6 grenade vests loaded with numerous flechette and fragmentation grenades. He also carried whole bags full of smoke grenades, frag grenades, CS grenades and Willie Pete (white phosphorus) grenades. He even had 2 thermite grenades for burning up the aircraft in case we were shot down. When we were flying over land he would always "trim" the treetops and if we were flying over water, he could get the windshield washed by the spray of the waves.

Now "you ain't gonna believe this":

On this particular mission I had to fly from the battery to Battalion HQs at Phu Hiep for a Battery Commander and Staff briefing in which we briefed the Bn CO and staff on our previous week's missions, any personnel or equipment losses we had sustained and any other "problems" we were having. Andy flew over to pick me up. He looked like Pancho Villa with his big handle-bar mustache and the slew of bandoliers of grenades he was wearing. We flew northeast to An Khe to pick up a part for our single-side band radio, then east to Qui Nhon on the coast to drop off a broken aiming device for one of my guns. We flew the route "nap-of-the-earth," meaning we were one with the terrain. Now the first time I flew with Andy, he gave me some rudimentary lessons on flying the aircraft in the event he was ever shot while flying. Of course, I was scared of the idea of going down, but I paid attention anyway. So, I flew in the left seat with my hands poised above the collective and the cyclic just in case.

We flew toward An Khe and went through the Mang Yang Pass, where in June of 1954, French Mobile Group 100 was decimated by the Viet Minh. In fact, that battle was depicted in the opening of the movie *We Were Soldiers* with Mel Gibson. We flew by the cemetery of white crosses that the local villagers created by burying all the French soldiers (some say they were buried standing-up facing west toward France).

We picked up the part for our radio, then hopped to Qui Nhon and dropped off the collimator, then to the airfield for fuel. We turned south and headed for Phu Hiep on the South China Sea. We were creating a rotor wash as we beat feet for HQs. Just north of

Phu Hiep is Tuy Hoa Air Base, home of a squadron of F105s, better known as "Thuds". The active runway ran west to east with the east end just above the beach. Whenever we flew near the runway Andy would always go as low as possible to clear the runway. As we're flying toward Tuy Hoa, Andy keys the mic and tells me he is going to request permission to "boom" the tower. Now if you've seen the movie *Top Gun,* you'll know that Maverick did a high-speed fly-by of the tower breaking the sound barrier and making a loud boom sound. But we're in an OH-6 helicopter and there is no way we can break the sound barrier. Anyway, Andy keys the mike and calls Tuy Hoa Air Traffic Control. Keep in mind that Phu Hiep also had an ARVN air detachment and had a Vietnamese air traffic controller in the Tuy Hoa tower. Andy asked the tower if we could "boom" the tower. He was granted permission, and we proceeded toward the tower. Just as we were adjacent to the tower, Andy pulled straight up and hovered even with the tower. He rotated the aircraft, so we were facing the controllers and keyed the mic. He then yelled "boom" into the mic. After which he dropped down to just above the beach and we hauled ass to the helipad at Bn HQs. As we were leaving Tuy Hoa airspace I could hear the controllers cussing in English and Vietnamese talking about some crazy Army aviator that looked like Pancho Villa. What a laugh!!

Anyway, that's my story and I'm sticking to it and it's up to you to decide if it was a fairy tale or a war story.

Dear Momma: Letter Home from Nam
Allan Perkal

This is a letter that couldn't be written then but can be written today.

Well, it seems like a long time ago that I was back in the world. Vietnam has been a journey into a land and time that I was not familiar with.

My unit, the 26th Casualty Staging Flight, was one of three casualty staging flights to take care of the wounded and evacuate them to destinations like the Philippines, Japan, and the United States.

My training as a medic supposedly prepared me for what I was going to do but, truth be told, as a twenty-year old, I would be facing challenges psychologically and physically that would later define "who I am."

The wounded came from all areas of Nam, North and South, East and West. Whether it was shrapnel, RPG rounds, or being burned, the wounded faced fears that they never had experienced before. They asked *why me*? *What was it all for*? *Who will live and who will die*? I didn't have the answers. My job was to keep them alive, help them cope with their suffering, and be their initial beacon of light on their road to recovery.

I do not remember their names, but I can see their faces and their wounds-especially the burn patients, the amputees, and those we kept alive on Bird Respirators.

I had noticed later in my tour a change was happening. I couldn't put a finger on it. I found the young kid that went to Vietnam had lost his innocence and aged well beyond his years!

The year in Vietnam, 1967-1968, taking care of those who borne the battle, defined me for the rest of my life.

Momma, you often wondered what happened to me. I was unable to answer that question. All you saw was an angry man. On my journey through the last fifty-four years I have discovered, like other Vietnam veterans, what was behind the anger, sadness, guilt, loss, and betrayal! Momma, I am now at peace with my Vietnam.

Rest in peace.

Your loving son,

Allan

Letter To the Enemy
Sarah Scully

Why didn't you do anything?
Anything to help us?
To protect us?
To save us?

We were your soldiers.
You were our Command Sergeant Major.
You saw their scared faces.

They were young females,
under your protection,
under your command,
begging for help,
begging for the simple right to take a shower
without fear of being raped by their fellow soldiers.

You told me it was the mission.
It was an exercise on the Korean DMZ.
They needed to suck it up.

I told you the male soldiers leered and grabbed at them.
The males in their unit told them
the females were there for sexual entertainment.
They looked to me for help

because I was a young female sergeant
traveling with you, the leader of all enlisted soldiers.

Me, with an illusion of power, trapped, just like them.
Them, with hope of aid, betrayed, by me and you.
You, with actual power as our boss, free, to help us all.

And you are saying, it's fine,
it's normal,
it's what they should deal with
and deserve for daring
to join the U.S. Army?

You sneered back at me
and said in a whispering, menacing tone,
"Females shouldn't even be in the Army.
Even then, they should have stayed in their places
as secretaries or nurses –
not as military police.
What next?
Combat roles?"

But ha,
joke's on him.

Every day in uniform,

every day as a female
in the U.S. Army
is a day in combat.

What I Carried
Ron Toler

    Yes, I carried the Smith and Wesson 38, the emergency radio set on Guard, the survival gear
with the chits for the villagers. I wore the nomex flight suit, the gloves, the ceramic helmet,
the jungle boots, and the flak jacket like all the pilots of that time and place.
    What I still carry is more telling, all the questions. What happened to you, Bill Finn, on that dark December night? Were you looking for bad guys on the ground and they found you first? They never found a trace of you, no matter how hard they tried. Should I have been there looking for you, too? You were the quiet one, did you open up once you were there?
    What happened to you, Dick Voigt, dropping sandwiches to the guys at that firebase from 25 feet, before your aircraft rolled inverted and crashed and burned. You were the wild one, always pressing the limits. I guess we weren't as invincible as we thought.
    What happened to you, Bobby Unrue? They said you drowned in the jungles of Vietnam! How did an athlete like you drown in the jungle? Was it a flash flood during the monsoon rains? You were the nice one, how did you end up there?
    I visited you three last weekend, I saw your names on that Black Wall and I asked you those questions, but no matter how hard I listened, you kept your secrets to yourself. I guess I'll have to wait a while longer until I'm there with you to finally hear your secrets and have the answers to my questions.

Transition to Limboland
Beth Angel

For many in the military, former service members and families, transition becomes a lifestyle.

From home fires burning to PT Rain or Shine, barracks with bunks to tents in trenches, struggling to survive in extremes as your toes turn black or you are sweating your balls off.

Briefings, Firewatches, non-potable water tanks, burnpits, monsoon downpours where leeches latch onto you as you muck through jungle swamps. Cleaning your weapon, bug out bags ready to roll, haircuts, shaving, uniforms ready for inspection, boots shined. You know the drill, as you count the days 'til you are short or prepare to deploy.

Uniforms, drills, training and more marching. Downtime is non-existent; hurry up and wait. Smoke 'em if you got 'em.
Chow lines, Mail Call, Sleeplessness, Hypervigilance, Trigger Happy and Zone Outs. Battle Buddies busting your chops, a girl in every port and itchy wool blankets on army cots. Sand or dirt in every crack.

Separation from Active Duty then brings that unforeseen shift. Safe from snipers, hazing, direct orders and C-Rats you now enter LimboLand. No one is telling you what to do, where to go; every hour no longer accounted for. You are free from duty, yet you yearn for those mad rushes, the missions, the chaos, the camaraderie.

Transition seems eternal as you strive to reenter a life that now seems utterly foreign; shell-shocked and ears still ringing from the flightline or frontline. Sink or Swim!

Gambit Prayer
Allen Utterback

"Let us pray.
Dear Lord, keep us safe
as we go back South.
Let the gunners aim be true
and watch over us
So we can go home too."
"Stop fucking around
and pay attention!"
Gambit pauses for a moment
then finishes: "Amen."
Amen in unison.
Sergeant Major Love
asks me about my smile
as we mount up.
"You cursed during Gambit's prayer
so we are getting blown
the fuck up tonight!"

Survivor's Guilt
Roy Moore

    I watch too many commercials; everyone in the military is having a great time serving the country. I'm going to be a career man; go air force! I drank the recruiter Kool-Aid—it's the best, goes down smooth for 18-year-old boys and girls. Basic training reality check Kool-Aid—damn what did I sign on for? Peace time Kool-Aid—just add gin, not a bad life, add more gin. Wow it just got real. Wartime Kool-Aid—what's that flavor? It's bitter, tastes like death and destruction, burning oil and soot. I'm going to need something stronger, I'm self-medicating out of this! Why did I think I could join the military and not get my hands dirty, bloody, swollen. I question everything I was brought up to believe in. All this pain from a damn commercial! I just wanted a better life. I'm from a poor black neighborhood, I just wanted out! And in return I got a lifetime of suffering, bad dreams, lost relationships, trust issues, self-medicating with the regular medication—how's that working out? Years later, my most vivid memory are those 9 caskets on the c-5 aircraft.

My thoughts then: damn, they didn't make it.

My thoughts now: they didn't miss anything.

Leaving El Toro
Jackie White

Early AM hours, transport fueled and filled
Band of soldiers, 84, this was not a drill
They had just taken off, mountain in their view
Nam they were headed, that was all they knew

No time for panic, 4 minutes in the air
No time to prepare, only time for prayer
Totally unexpected, tragic and unfair
If only this had been just a nightmare

There were no survivors that day
On a grassy Loma Linda hillside
This is where their ghosts remain
Under the fog and misty rain

It took four hours to reach the site
8000 gallons of fuel burned that night
A deadly stillness remains in that air
Families mourning with the pain to bare

No one was left behind that day
Forever together in this disastrous way
Bound by the courage they each displayed
Unanswered questions still remain

They missed the homes and folks below
Time stood still, that they know
And then the silence of despair
No more together, no more there

They rose above the fear that day
Into darkness with no delay

They chose the mountain straight ahead
So others would not die instead

It was June 25th of the 1965th year
There was no time even for a tear
The dust and sirens no more roar
This was the day the tribe did soar

"El Toro" is no more today
The Great Park is now displayed
Finally after decades, a plaque erected
No cause for crash ever detected

       Dedicated to L/Cpl John G Brusso, Ontario NY

Letter to That Artillery Crew
Wallace Bohanan

At first, we just sat atop our bunkers and watched as your artillery crew zeroed in on our Firebase. The problem was our bunkers weren't built to withstand the rounds you sent roaring our way. Even worse, your gun had a range that only our 175mm cannons could reach. Circumstances worsened when my gun crew was chosen to respond. That meant that everyone on the Firebase was hunkered down while we alone loaded up, raised the barrel, and fired. As soon as we sent our round downrange, yours came screeching in, hot and explosive. We waited out in the open until the report came that we had scored a direct hit. Our rounds must have crossed paths in the air.

I had been so shaken by your incoming artillery that I was both relieved and elated that we had eliminated your crew. That was then. When I think of you now, there is no relief, and the elation is long gone. I realize we were trying to kill each other because we were ordered to. We were soldiers in circumstances beyond our control. I didn't want to be there, and I didn't want to fight. Yet I did fight, and you died as a result of my actions. I don't ask for forgiveness. No shame, no blame. However, I am profoundly sorry for the death of your gun crew. There were so many lost lives and grieving loved ones on both sides of the battlefield.

Artillery crewmen usually don't know who their rounds killed and maimed. In this case, I am sincerely saddened for my part in the actions that caused the loss of your life.

You Saw Me but You Didn't
Beth Angel

You saw me in my uniform, with that frozen smile upon my face
Looking official in my SP beret, masking rage, leaving no trace.

Working a dog overseas, one female on a 20-man team
Mandated an identity shift, my reactions harsh, even extreme.

Family and friends back home had absolutely no clue
What it took to survive rape, hazing, threats, brutality, too.

I don't resent those enlisted women left unscarred and untouched
But the judgement or disbelief of my experiences was simply too much.

I've encountered many military sisters who also survived those things I overcame
It outrages us to discover young female soldiers now serving are treated the same.

There's no way to prepare you for what occurred during my tour of duty

When asked if I'd serve again, I respond,

        "Fuckin'A, Damn Tootie!"

Who Am I
Allan Perkal

I am a 78-year-old Philly Boy, entering the fourth quarter of my life, who lives in the Mountains of Western North Carolina.

I am a member of a Tribe called Brothers and Sisters Like These.

I am dedicated to making a difference in the lives of those who served their country in war.

I am first generation US Citizen of Polish parents who came to this country to find freedom from persecution.

I am a medic who served in the Vietnam War that defined what his life was to be!

I am a loving husband, stepfather, and grandfather, who has been blessed being part of their lives.

I am a human being who believes in a cause greater than myself.

I am a passionate person who will meet his maker knowing I did my best to make a difference in this world.

Reflection
Stephen Henderson

Did I do enough? How many died after I left them? They were all great Marines.

What Saves Us
Roy Moore

What saves us? We are the latest version of the roman empire; we rule the world, they hate us. But we have the technology, we have the respect of the world. It helps to have the strongest military in the world. That brings respect and fear; there's no bear we won't poke—that saves us! What saves us? Arrogance, confidence, Americans like a good fight. We're cowboys, we're the heavyweight champions of the world. When we put our mind to it, we are unstoppable and that saves us! We have a common thread: "don't tread on me." Remember the towers? Someone just poked us. Big mistake. We still didn't pull out our A-game. That saves us. What saves us? My training, my boys in this bunker with me, trust that we are doing the right thing (that's a question for later), morals, character, my family, your family. Black, white, Asian, Latino, sometimes we're all on the same team. That's what saves us!

Veterans Day USA
Ron Kuebler

Land, sea, air of USA are defended by Americans in uniform
Veterans all but many paid the maximum price in the storm
We owe them our utmost respect as they died in defense
Of our beloved America leaving their comrades to make sense
Of a country defended to the death for some
Surviving veterans bear the scars of war and live for the freedom
We all treasure so much to the deepest depths of our hearts
Thanks to their sacrifice for all of us so we can eat the tarts
Of apple, enjoy freedom in a free country and savor our lives
And express our freedom to others so that they may thrive
And bear children who know freedom and what it costs
They, in turn, carry the banner of self-sacrifice maybe to loss
All for the freedom we hold so close to our hearts.

Wife of a Vietnam Veteran: Married to the War
Marsha Lee Baker

Tom joined the Army and arrived in Vietnam in 1968. He was 19 years old, fresh out of high school in Jackson County, North Carolina. I was 14 years old heading into high school, living with my parents and brother on an Air Force base in Florida. Tom and I met 35 years later at Western Carolina University and have now celebrated our 21st anniversary.

It never occurred to me at the time that I was a military daughter in love with a military husband. It never occurred to our local friends that we would become a match. He's such a war guy, and she's such a peace gal! Yet, Tom and I saw our union as "meant to be" and had many conversations sharing our perspectives about war and peace, violence and nonviolence, local and global situations. We enjoyed learning from each other's perspectives. We didn't have quarrels; we had conversations.

On September 11, 2001, when our country was attacked by al-Qaeda terrorists, we talked more than ever, particularly about what our country's response should be. It was a serious, intense conversation, and we knew our country and its people were also struggling.

As usual, Tom talked about history, wars and the military, and I talked about violence, nonviolence, and peace as we imagined our country's response. There wasn't a lot of common ground, though a lot of listening, with our love and respect for each other carried the day.

Over time, Tom began to share with me a few of his most difficult memories from Vietnam. He wept one night as he told me about shooting two Vietnamese farmers. He was a helicopter doorgunner, and the chopper was cruising over a field. It was designated as a "free kill zone" by the Army as part of an effort to increase their number of kills. He wept as he told me with grief and remorse that he had shot two Vietnamese farmers and saw them fall dead as the chopper moved on. "God Damn! What had they done to deserve that? I hate living with what I did!"

He shares a story in his memoir about risking his own life to save the life of a small Vietnamese girl. He also tells it when he is invited as a veteran to give a talk, and his audience weeps with him.

Tom always enjoyed telling tales about the camaraderie shared among his closest military buddies, and how he and Patrick O'Shaughnessy were best friends. They were later deployed to separate locations in Nam. I think Tom has never forgiven himself for somehow, someway, not being able to save O'Shaughnessy's life. He has visited The Wall several times, each time touching his friend's name and weeping.

Once when I returned from a long academic conference, Tom somberly welcomed me home. "I don't do so well when I'm home without you." Not imagining at all what he meant, I replied, "Well, honey, I'm going to be gone off and on." Beyond clueless, I was unthoughtful. As I listened, he told me that he had been watching war movies and drinking Bourbon. Then he went upstairs to his man cave, which is filled with war memories and memorabilia. He pulled out his pistol and kept thinking. Then he heard some of his veteran friends talking to him, telling him he still had much to do. I don't know how long he sat there contemplating what to do or not do with his pistol, though he eventually listened to the voices and put down the pistol.

I was dumbfounded, speechless, frightened. I fell into his arms, and we cried. I attempted to say how thankful I was that he was alive and told me about this experience. We each in our own way relive that story and its possible endings time and again. Once more recently, he sunk into a deep dark hole, and I removed the pistol from home until he was back in the care of his VA counselor.

Within the past decade, I have begun to pay closer attention to how Tom's health was and is affected by his surroundings. He does not always sleep soundly and often wakes up with, or later recalls, nightmares. The national and global news troubles his mind and disposition. He habitually unloads profane anger on drivers. Unexpected sounds that otherwise are familiar in war heighten his anxiety—fireworks unannounced or similar sounds that he cannot immediately locate, large crowds anywhere from outdoor events to indoor concerts, and being in crowds while shopping or at a restaurant. I was beginning to recognize PTSD without knowing it.

In about 2002, Tom realized that he had a story to tell, and more since. I was a professor in rhetoric and writing, so he began asking me questions that many writers have as they begin to write with serious intent. We developed a two-fold partnership as writer and reader as well as war veteran and writing professor.

DAUGHTER OF AIR FORCE C-130 Gunship Pilot

My Dad was a career officer for twenty years, and we moved around a lot. I was born in North Carolina on a military base, and my brother, my only sibling, was born in Ohio four and a half years later at yet another Air Force base. Dad was gung-ho about flying in the military. In later years, he told me that he had wanted to be a fighter pilot. I asked him why he wasn't, and he quietly answered, "Your Mom didn't want me to because she thought it was too dangerous."

I realized years later that my Mom, as a military wife, had most of the responsibility for raising us and making sure we behaved as respectful military family. Proper manners, appropriate clothing, follow the rules, do well in school, and do not get into any trouble. When I look back at family photos and memorabilia, I also see pressures on the "wife" to dress and socialize as Officers Wives followed a protocol all their own. She shared with me years later I that she was especially nervous about what she perceived as her limited abilities as a young woman from Greenville, SC who also did not complete her high school education.

John and I were confused during these military decades about how and, moreover, why our parents argued as harshly as they did yelling, hitting, leaving, and alcohol drinking. In their era, I doubt any counseling was sought or available. Even after their marriage settled down in retirement, our mother continued to call each of us to question and criticize why we were doing or not doing what she thought was best.

My parents lived long enough for them to know Tom and us as a couple. They liked him, trusted and enjoyed him. It was a special and precious time. He and my dad had conversations sharing their good and bad military experiences. They laughed a lot and drank beers together. The treasure for John and me is that Tom told us details about what Dad did and what he said about his experiences . . . stories and memories we have never known and could not find out.

Tom was there when Dad passed. He was there when Mom passed.

Thank you, Tom Baker for asking me to marry you. You're welcome for me saying Yes!

4th of July: Hot off the Grill
Richard Epstein

The front yard was covered with cars and pickup trucks.
Bumper stickers gave it away: Combat Vet, USMC,
Purple Heart, Attack by Air, and Beaucoup Dinkey Dow.
Tables and chairs huddled in front of an empty garage.
Smiling old men-- some short, some tall, some portly,
some thin, gesture wildly as they sit over their food.

Uncle Than, tongs in hand, watches over a hot gas grill
covered with hot dogs, hamburgers, and large shrimp.
No weapons in sight. No Ba Muoi Ba, no Tiger Beer, no grimy
boots and worn fatigues, no tall bamboo, no odor of nuoc mam.

The men talk about how hot it was at the parade today,
a recent skirmish at the VA, a trip overseas,
a summer cruise, the plan to move-in with the kids.

Doc, a medic, looks down at his plate and quietly says:
"The guy I replaced didn't make it. The poor bastard
who replaced me didn't make it either. You know,
I still have shrapnel in my legs."

Someone asks: "Where were you, Doc, I Corps?"
"II Corps—Central Highlands with the little people:
Yards we called 'em." Then several guys talk about
where they were, what unit they were in.

A heavy-set man with a ponytail shakes his head and says:
"We were younger then. Hell, most of us were just boys.
I was Green Beret, SOG. I blew up stuff. Damn glad I served.
But I'm happy I made it back home."

He raises his glass and says: "Here's to Doc
and the Yards, and here's to those who got left behind."
"Here, Here," they shout. "Here, Here."

Letter to the Enemy – I Haven't Thought About You for a Long Time
Donna Culp

Dear Enemy,

It feels strange writing this letter to you. Mind you, I wouldn't have given much thought to write to you except that I'm in this veteran writing group and this is our assignment. I guess I just thought that since we were on opposite sides, we had our respective jobs to do, and that is that. Given the reason I'm writing this to you, there must be some value in the exercise of dredging up thoughts I probably have simply, or not so simply, stuffed into a corner of my subconscious. The crazy thing is that both of us are flesh and blood. Just regular human beings. Thoughts, feelings, families, friends, homes, communities, and well, just humans at the end of the day.

Over the years, the opportunities to meet have been many. Sometimes we met face to face, and other times I could only sense your presence. Knowing you were lurking in the shadows, measuring the best opportunity to muster all your predatory skill and pounce. It is the lurking and measuring I despise most. When you fell short of your mission, I counted that as a victory for my side.

I must be a HVT (High Value Target) for sure, because you see some value in killing me or morally destroying me for some sinister and self-serving reason. Is it ideological? Is it insecurity? Is it pridefulness? Is it insanity? Is there something I have that you so desperately want that you'll stop at nothing to get it?

You have a reputation for culling out the perceived threat. I see your greed, and lust for power, and the thrill of overcoming your enemies through the most deceitful means possible. You set traps, pose as a false ally, and play to the tune of your leader for the carrot at the end of the stick. Sorry, dear enemy. I hope you choke on all the carrots you have won. Your day of reckoning is lurking just around the corner.

Where I'm From
Emiliano Enea

Where I'm from matters no more. Instead, *what* I'm from is what matters. My existence is but a mere line in a book; a book among millions in a library collecting dust, accessible only to the second-row shelf via a rolling ladder. A ladder that has been used only once to place the book in its placeholder and never to be used again.

What I'm from is the same thing that you came from, the most basic forms of life. The Sun, Water, Earth and Air. A ball of consolidated energy, denser than anything the Universe has ever been able to compartmentalize in the fabric of time and space. All of humanity laid within that core and with its explosion came the Universe we now know. Who or what created that energy remains academic to me, for what is important are the implications of what laid inside that core, for I come from the same place you came from. You are my Brother, you are my Sister. You are the animal that nourishes me, you are the plant that feeds me. I come from the air you breathe out and you come from the air I breathe in. We are one. I'm from the earth that has nourished those before me and to the earth shall I return to continue to nourish the future generations. And yet, we met on the battlefield, where I tried to kill you, and you tried to kill me.

I'm from the same place that should unite us all. A place that allows us to flourish and thrive, that gives us a sense of purpose, the ability to create a family and to leave a legacy, no matter how small. The things we have in common are much more than those that separate us. I am from you, and you are from me.

That is where I'm from…

***Where I'm From* Background**

The essence of the meaning behind this writing is to express the change in my spiritual life after my combat service. The physical, mental and emotional stress of living and dying in a prolonged combat environment, in both Afghanistan and twice in Iraq has changed my perspective on life. Perhaps, it is a way for me to make amends with my military experiences and cope with the struggles that plague many combat Veterans. The use of military intervention is the definition of political failure, and it is this sense of failed political and military institutions that have made me move away from the religious institutions and find peace in the notion that, in the end, we are all the same, we are all one.

What Saves Us
Allen Utterback

What saves us, I don't know?

It might be as individual

as everyone seems to be

peel off for a revealing.

The child knows no fear nor hate;

just a new clean slate.

Information obtained through senses during days.

Is it pencil and paper; erasable?

Or a stone tablet etched forever?

The child grows into a man

making choices based on…

I don't know.

What saves us is an unanswerable question.

Meat and Three
Charles R. Duke

"Whatcha want, hon? Beef or chicken?" A harried middle-aged waitress at an unairconditioned blue-plate-special diner at a crossroads village kindly waited a couple of seconds for my synapses to connect as I melted into a swivel stool at the worn Formica counter during the Great Plains August wheat harvest in the late '70's. My mind drifted for a couple of seconds over a massive amount of loosely connected experiences that influence these little decisions.

Early morning light shafts pierced old barn wood cracks on a kid's trudge to daybreak chores. Lights kept our 4-H chicks eating all night. Their mindless chaos scattered food through upturned water troughs. No attachment formed for a big flock of undifferentiated blobs of poultry. We chopped heads, scalded feathers, and gutted without remorse. But Wendy's tears stuck with me. Our milk cow's yearling was shot and butchered in front of her. She stopped giving milk for a month, bellowing constantly, so much that her irritated throat almost closed. For us, the yearling was just food.

During our nightly patrols over the Ho Chi Minh Trail, our sensors tracked lines of infrared heat headed south on the right side of the rules of engagement. We couldn't see the vegetation or the outlines, just the heat. If the line stopped when we began to fire, the blobs were trucks. If the blobs scattered off the trail in sporadic direction, they were elephants and buffaloes burdened with fuel and ammunition. It didn't take much to blow up the targets, shredding the forests all around with thousands of rounds per minute of 7.62 mm and 20 mm ammunition, just like the bullets shredded the people targets in close-air support for friendly troops, whether there were a hundred, 400, or 1000 enemy. Body counts were impossible even though the US public wanted to hear that our kills were more than our losses. Later we learned that our soldiers were traumatized by the sight of what was left of their foes after our attacks. Cows, calves, and bulls shed Wendy's tears for their losses. Like our tears for our buddies. Like the enemies' tears for their dead.

"Com'on, hon, I need to serve other folks," the sweaty hard-working lady prodded for a food decision.

"Sorry, ma'am," I choked, shocked to get back to reality. I kept my head down before I wiped my eyes. "I'll have chicken."

Letter to my Enemy
Roy Moore

I apologize for the scare, but I don't wake up well. I hope yours is as bad as mine. Since 1991 my life has been hell, no sleep, self-medicating, loss of family and friends, lack of compassion, people looking at me wondering what the hell is wrong with him. And you gave me this going away gift and I hope you got mine, but I'm on the winning side. Hey, my enemy, what does losing feel like? No sleep, self-medicating, loss of family and friends, lack of compassion, people looking at you wondering, *what is wrong with him?* I know you didn't start this, but I helped finish it. You're welcome. Rodney King said in March of 1991, "can't we all just get along?" My thoughts on this: apparently not! There's good in men and bad. Back then, I hated you. I don't anymore, I just dislike you now. You wanted me dead; I wanted you dead! No love lost. It's okay, you keep your gifts, I'll keep mine. I still don't sleep well, self-medicating— it stopped working after the 3rd or 4th body in my presence. My family is back in my life, I have new friends, compassion is coming back, people don't quite look at me the same. I apologize for the scare, but I don't sleep well.

Suicide: What is it Like
Ron Kuebler

What a word for a state of mind
We do not want to encourage
But weak we are to resist
Its hold on our psyche
Weak to leave its tenacious clutches
Strong to gather the tools of destruction
That tear apart our body and our mind
But first our body as we see death
Approaching relentlessly or quickly
So that our mental torment
Is forever gone away
Our memories that taunted us
Drained into the oblivion of nothingness.

When I knew I was Home
Donna Culp

Six months after returning to CONUS from a year in Iraq, Christmas arrived. This morning, we woke up to it snowing. Beautiful, huge, fluffy snowflakes gently falling from the sky, gradually accumulating on the ground. What a treat after a year of dirty brown surroundings, sandstorms, scorching temperatures, and rocket attacks.

Later that evening our dog, Molly, and I gingerly stepped out onto the front porch and then to the snow-covered sidewalk for our nightly stroll around the neighborhood. As we walked, it dawned on me how silent the night was with the snow muffling the sound of any ambient noise in the area. The most audible sounds were the jingle of Molly's metal tags on her collar and her breath as she sniffed the air. As we continued our walk, the Christmas lights decorating our neighborhood displayed a radiant but soft glow from underneath the blanket of snow. Then it occurred to me that here in this moment there was peace on this patch of earth, and goodwill toward all in our little corner of the world.

Last Christmas in Baghdad, our team members all gathered together with the awareness that in a heartbeat our little bit of peace and goodwill toward each other could be shattered. Realizing the stark difference between last Christmas and this one, I suddenly felt overwhelmed with a swirl of feelings ranging from gratitude to grief. Sensing a change in my mood, Molly stopped, and the crunch of snow beneath my feet stopped too. Molly looked up at me and nuzzled her nose up against my leg while simultaneously letting out a slight whimper. This sweet gesture was meant to let me know she was there for me. She had my back. Kneeling down next to her, I buried my face into the soft fur of her neck and just sobbed. Between sobs and feeling the weight of something lift that I didn't even know was there, I could hear my voice whispering a prayer asking God to help us find our way to peace on earth and goodwill toward all. I don't know how long this pause lasted, but I know we were there until the sobbing stopped. Molly and I eventually resumed our walk back home with a refreshed assurance that we were in a safe place.

In the back of my mind, I was holding our teammates in Iraq close to my heart, silently whispering a prayer for their safety and for a peaceful Christmas. That's when I knew I was home.

Memorial Day, 2019
Stephen Henderson

    Good morning! On this Memorial Day, across this great land, we pause to remember all men and women who have given the ultimate sacrifice to preserve the freedoms of the United States of America. You are here today because of the loss you may share with me… the loss of a spouse, a brother, a mother, a grandfather, a wonderful friend. Many of you may be here today because you have felt robbed of the relationship you lost with these important family members and friends. The questions of *what if?* and *if only*? you may have asked many times. You may be here today to remember the people they were and the love you shared.

    Today I want to remember and share with you the story of my friend, Gene Parton. Gene and I attended Woodfin School and Erwin High School together. Early in life, you knew that Gene had your back. We played together as boys, riding our bicycles all over the Erwin district to meet up with other friends to hang out for the day. We would ride the Elk Mountain bus to downtown Asheville, going to the movies at the Imperial or Plaza theatre. We would then walk home down Broadway to Riverside drive as the bus had stopped running. If we had any trouble, Gene was right there to make a plan or to make it right.

    In high school we played on the Erwin High school football team together, pushing forward each Friday night with the determination to win. Even when he was injured, he would never quit or not play. He gave 100%…he was tough as nails. Even in high school Gene Parton knew that in order to gain respect, one must give respect. As a student and athlete, he recognized the proper way to treat people and to help others.

    Some of you may remember the Tunnel Road of the late 60's… Buck's, Babe's, Winks, the Dreamland Drive-in, Shoney's… Gene was the center of attention in a positive way. On Patton Avenue, Erwin students hung out at Burger Chef and Enka students were at Surfside. We often met up at Eddie Joyner's speed shop. On several occasions Gene and I double dated, and we all had great fun. Once, Gene drove his love sick friend all the way to California to see his girlfriend…it was just the kind of guy Gene was.

After high school, Gene joined the US Marine Corps. This was a big deal to Gene as it was something he was doing for the first time on his own. As I recall Gene was the honor man in his platoon. This was his first of many awards to soon be earned. He completed boot camp at Parris Island and then completed advanced infantry training at Camp Geiger. Gene came home on a 30 day leave before completing survival training in California and the on to the war in Vietnam.

While he was home on leave, he spent a lot of time with his friends. We gave him a going away party. The last time I saw Gene, we were sitting in my Camaro listening to the "House of the Rising Sun" by the Animals. He shared with me stories about his training and his thoughts about going to war and serving with his fellow Marines. He always thought about others. I knew he would be missed, not knowing that I would never see him alive again.

While in Vietnam, Gene was a selfless warrior. His actions and awards were that of a hero. Less than a year later, his body was returned to Asheville, a flag draped over his casket. The hurt of his family so raw, the loss of Gene was felt by the entire community. Never an ordinary guy, Gene represented the guts and glory of a true hero, one who affected our lives in a never-ending way

Bobby Horton went through boot camp and to war with Gene. He said, "He told my wife that he would take care of me and get me back home to her, but at that time I thought we would both make it home. Gene was the brother I never had and I think of him often. I named my first son after him and my Dad. It has been 35 years but sometimes it only seems like yesterday we were talking, laughing, and looking to the future. He will never be forgotten."

The Silver Star was awarded for actions during Vietnam to Gene Parton:

The President of the United States of America takes pride in presenting the Silver Star (Posthumously) to Lance Corporal Floyd E Parton, United States Marine Corps, for conspicuous gallantry and intrepidity in action while serving as a fire team leader with Company B, first Battalion, fourth Marines, Third Marine Division. In connection with combat operations against the enemy in the Republic of Vietnam. During the early morning hours of 8 October 1968, while his platoon was occupying a night defensive position near Khe Sanh, Lance Corporal Parton was on the perimeter security watch when he detected movement to his immediate front. Observing an estimated squad of North Vietnamese soldiers, he was attempting to warn his team when the enemy delivered heavy fire. With complete disregard of his own safety, Lance Corporal Parton immediately threw numerous hand grenades at the enemy soldiers and delivered accurate suppressive fire at them in a fearless attempt of halt the assault. As he moved about the fire swept area, shouting words of encouragement to his men and directing their fire, Lance Corporal Parton was severely wounded. Undaunted by his serious injury, he valiantly continued to fire against the enemy and encourage his men until he succumbed to his wound. His aggressive fighting spirit and determination inspired all who observed him and were instrumental in the defeat of the hostile force. By his courage, bold initiative and unwavering devotion to duty, Lnc Corporal Parton upheld the highest traditions of the Marine Corp and the United States Naval Service. He gallantly gave his life to his country.

May we never forget those who have died while serving our country.

Survivor Guilt
Dorian Dula

The 1st Battalion of the 5th Marine Regiment of the 1st Marine Division deployed to Vietnam in June of 1966 and returned to the states in March of 1971. During that time, 142 Marines of Charlie Co. were KIA. That's just one company of that Battalion. During my tour from 4/67 – 2/68, 53 Marines were killed. 3 men were KIA and 9 Marines were WIA in the squad I was in charge of on the morning of 2/18/68 in the Battle for Hue City in the Tet Offensive of 1968. I saw a grenade roll into the room we were in and yelled to everyone, "a grenade!" Boice ran out of the house and I was right behind him. The NVA then opened up on us and Boice died instantly. I received a round in my upper right thigh. The grenade went off and everyone left in the house were WIA but Berry and Warren were KIA.

I have replayed that morning in my mind thousands of times. Why did Boice die, and I didn't? Why did Berry and Warren die, and I didn't? I was in charge; what if I would have done *this*? What if I would have done *that*, maybe things would have been different. I've thought of the events of that fateful morning thousands and thousands of times. Why did I make it home and my name isn't on the wall with the other 58,000+? I've been in therapy off and on since 1970, I'm not ashamed to admit that I've cried countless times about the war. I've had some severe PTSD sessions since I came home. Mostly just severe depression that can last for days. I'm rated 50% VA disability for PTSD and 40% for my physical wounds. My physical wounds have healed but my mental ones never will. I called the Suicide Hotline one night in August 2021. I needed someone to talk to, I was so down.

On that same morning in Hue City, I thought I heard movement in one of the rooms in the house. It was still dark as it was dawn. I sprayed the room with all the rounds in my magazine. As I was loading another magazine into my M-16 I saw there was an open door leading outside. In the shadows I saw a gook with his rifle pointed right at me. I thought I was dead but he ran away. I cheated death twice in an hour. Is there a God? Was he looking after me? If he was, why didn't he look after the other guys on the wall?

In the room I'm in right now, on the wall, is a nice, framed manuscript. It has the name, rank, city of origin and date of death of all 142 Marines I mentioned previously. I'll never forget all the friends I lost as well as other Marines I barely knew that didn't make it. Everyone in this room will never be the same for what we went through. The Vietnam war ended in 1975 but many of us are still fighting the war.

Standing In Front of You
Beth Angel

Standing in front of you, bearing my soul was not on my To-Do List. Compelled to continue my "process," the Healing Journey meant crossing the lines, volunteering to deal with that old boy bullshit. JUMP IN and just, put it out there.

Female Veterans are the same, but different.
Warriors. Soldiers. Sisters in Arms. We're a hard-headed badass bunch. We drop "F" Bombs and hail you bitches with Texas salutes. All-Day-Long.

Gentle-Down, My Ass.
We lose our Amazon Identities when treated like *Dependos*, not recognized for our service because we are female veterans. We slip into married-with-children-life as seasons of soccer and unicorn princesses replace ironed uniforms, polished boots, and a side arm. We fit in no mainstream categories; civilian labels don't correspond to the identities or roles that defined us in the military. That way of life is no more; we feel the vacuum and lack of mission essentialness as do our male counterparts.

I'm mission-driven to focus on a way for vets to keep their shit together on the outside. It took way too many years to learn. It's not rocket science; it's called CAM-ER-AD-ER-IE.

You may be an ancient old warrior, a shunned 'Nam vet, a non-com rated troop or maybe you survived the insanity of the Sandbox. Your MOS didn't define you as much as your patriotism did.

You may feel unworthy because you didn't wind up downrange or get blown up. It's a NUMBERS GAME; we ALL signed our rights away. We all had our brains scrambled, our racks flipped whether Boot Camp or Basic Training. That banging of trash can lids crazier than clanging cymbals, the hollering, the lack of sleep and privacy. Sleep deprivation, mind games and physical torture are our common grounds. Giving up our rights was expected.

Even peacetime assignments can incur injury, trauma or death. It's not a fuckin' competition. It's a Brotherhood with some sisters thrown in to shake things up. We all signed up for this.

IT TOOK DECADES to force me to unpack my old "trunks." Then came the Forced Rest.
Stashing shit away, shoving it down and out, never resting, always working to keep my mind at a dull roar.

No services for female veterans means you jump into a men's depression (or whatever is available) group. Those blue man group eyebrow's raise in unison when an old warhorse opens her mouth. Learning that a female veteran's depression manifests in RAGE came as a shock to all; it was funny until it was not.

Why does that healing process journey take so damn long?

Too many of us old, disgruntled vets hit that wall – often – and pull back to lick our wounds. Everything hurts. We become tired of standing in lines. Appointments, Evals, Consults, tests, labs, X-rays, but no cure. We don't want to have to ask for help. The system is exhausting. We grin and bear it. We take a lickin' and keep on tickin'. No Guts – No Glory.
But – DID YOU DIE?!

STANDING HERE BEFORE YOU- I plead with every one of you who served, FIND SOME BATTLES! DO WHAT HELPS & DO WHAT YOU LOVE.
As you invest in your sanity, recovery, restoration – you may find SIMPLICITY SUITS YOU.
Find a Battle Buddy or two. Text - Call: Buddy Check. Do random meetups or attend shit together.
Here's the sales pitch: Journal. Write. Listen to music. Play an instrument. Paint. Draw or color. Beware if your battles are Jarheads because Marines eat all the crayons.

Don't shelf your shit. Truth hurts but heals.
You need a minute to unpack all that baggage.
Transition with intention. Be kind to your own self.
Don't recreate the wheel - find folks that are doing your thing- and just jump In.

LIVE FOR THOSE WHO CAN'T.

Eight Green Thumbtacks
Tom Hickey

"You ever kill anyone?" he asked.

I froze and stared straight ahead. Nobody asks this question. But this was a fellow vet as well as a patient who I have a moral obligation to be straight with. He considered my one-word answer. Then he said, "Well I killed someone," and told me his story.

At a Fire Support Base in Vietnam, he was firing his 105 howitzer fast and furious over the heads of a Marine unit in danger of being overrun. There were so many shells in the air that the whole area was declared a no-fly zone. Just as he yanked again on the firing lanyard, to his horror a helicopter rose straight up from among the besieged Marines. He watched helplessly as his shell spiraled right into the big red cross on the side of the chopper. It blew up in a huge ball of flame with big burning pieces tumbling to the ground.

For the next fifty years he only slept a few hours a night. Obsessed with the men he had killed, he endlessly researched the casualty lists and units involved in that day's action. Every time he came to see me at the VA, no matter if the appointment was for his blood pressure or other issues, he always turned the conversation to "The Helicopter."

One day I was driving to work, I wasn't thinking about the patients waiting for me because I never knew who was there until I got there. I was wondering when my Amazon package would arrive. Then I wondered, what is in that Amazon package? I realized that if I can't remember what I ordered, do I really need it? As I wondered, *what's really important in life?* my eyes started drifting upward, like the answer might be "up there" somewhere. When I got to work, the first patient on my list was the Helicopter Guy. (Probably not a coincidence.) We were supposed to talk about his pills but quickly he went to The Helicopter. But today I was suddenly inspired. I dug out a can of multicolored thumbtacks. I picked out eight green thumbtacks and laid them on my desk.

"These are the eight guys you killed," I said.

"Yes, yes!" he said. "I can see their faces…!"

"Okay okay," I said. Then I put a couple of dozen more tacks on my desk and said, "These are the guys whose lives you saved by killing all those VC who were trying to kill the Marines."

He looked back and forth between the two piles of thumbtack Marines, then said hesitantly, "But I still killed these guys," pointing at the eight green thumbtacks.

"Yes, true," I said. Then I dumped out all the remaining thumbtacks on my desk. "These," I said, "are the children and grandchildren of those Marines you saved, because you killed the guys that were trying to kill them. So, they got to go home and live their lives and raise families."

He stared for a while. Then he said, "I never thought of it like that."

Whoever saves a single life, it is as if he saved the entire world.

(The Talmud)

Farm That Heals
Ron Kuebler

Veterans make this a healing farm

Military service provides wounds that harm

Way after the time-of-service past

The wounds seem to fester and last

Farming distracts the mind from self

Growing for others you can be an elf

Sweet potatoes, tomatoes, blueberries and more

Giving to others is not too much of a chore

Makes you feel better about your life

Storing in a corner goes all that strife

Restoring your mind to a better place

Healing Farms may be your space.

Aftermath
Charles R. Duke

American culture evolves and shifts on a pendulum swing: conservative to liberal; ethnocentric to worldview; polarized to unified. A dozen years of the Vietnam war pushed us away from unity with an equitable draft policy by creating privileged exclusions. The conflicts at home created deep divides in attitudes toward the military and those who served.

To set the atmosphere, I was there during the last year of the war. I came home on leave for my wedding and then back to Vietnam. It was financial efficiency, you see. On the way to my first post-Vietnam assignment after the cease fire in 1973, my new wife's highly educated civilian friends hardly looked at me during her going-away gathering. After a lot of wine and beer, one fellow admitted confidentially, "they just don't want to ask her why she married somebody who went to war instead of someone who went to Canada." I retreated into the background, to avoid the conflicts.

By 1979, I was applying for a major promotion within my company and the manager asked me, "are you sure you can handle this corporate demand for precision, I mean, since you were in the Vietnam military?" Losing the war made all former military seem incompetent. Society had gone from "how could you fight in that war?" to "how could you be the first to completely lose a war for America?" I answered, "When you're in it, you do your best. Someone said that the military's job is to go where no one should have to go, see what no one should have to see, and do what no one should have to do. Then, if you're lucky, you get to go home." He just smirked, but I got the job.

Why join and go to war was an easy answer. Southerners at the time were "over-represented" in the military. For us, respect for authority and obligation to serve were strong motivators for being there, along with the leftover image of honor from the War Between the States. William Faulkner's description in "Intruder in the Dust" distilled the feelings into a great image:

For every southern boy, it's always within his reach to imagine it being one o'clock on an early July day in 1863. The guns are laid, the troops are lined up, the flags are out of their cases and ready to be unfurled. But it hasn't happened yet. And he can go back in his mind to the time before the war was going to be lost, and he can always have that moment for himself.

"Hush, honey, don't say anything to Uncle Charlie about Vietnam," chastised my sister-in-law when my young niece asked about a new war movie being shown. By the mid-1980's, society had gone from "we're embarrassed that you lost the war" to "we won't mention it because you might go berserk." I retreated again into myself not wanting to create conflict in the family.

"I didn't know you were in the military," smiled just about everyone after the 100-hour first Iraq war restored some of the public's confidence. The news coverage of the bad parts of the conflicts reminded me of the destruction we meted out in the jungles. "I don't know why we were there or what good it did," I admitted to guys at a college town bar in the mid-90's. They truly wanted to understand. One of the history experts replied, "have you read McNamara's book? The old codger really gutted the US's decision making." So, I read the book and watched the multi hour television special interviewing him. I told my wife, "Now I see the reasons and the fallacies." She observed, "It's been 20 years, nobody in the public cares." All I could say was "I care."

"Thank you for your service," say the people around us today as they quickly smile, then turn away. Twenty years of desert wars have raised awareness and appreciation. After a few beers, a psych expert surmised, "They have no way to understand what veterans have been through, so they just run away from their own inability to comprehend." I said, "Yeah, but I have the experience, and I just don't want to make vets talk about it."

Then I was doing a Veterans Day program and found a great approach to expressing appreciation. "Thank you." Two words, eight letters. We say it all the time often without thinking. So, where is the power and the emotion? When do two words earn their stripes as true gratitude and appreciation?

It happens when we look into someone's eyes; when we remember what we said and to whom we said it because we mean it. It happens when we don't plan it and aren't prepared for it, but we are driven to say it. When we let our hearts speak. Maybe we could say, "I don't know how you did what you did. But you had to do it. And I thank you for your sacrifice."

It happens when we listen, really listen, especially in-between the sentences, and behind the words. Listen to the emotions being pushed down and the fears that are walled up. Listen to the soul that has been scarred with the places no one is supposed to go, from the sights that no one is supposed to see, and from the things that no one is supposed to do.

Just listen.

What is a Veteran
Ted Minnick

Many times, when someone says, "thank you for your service," I will ask them if they served. Some have said they served but didn't see combat or they served but never got to Vietnam, calling themselves "Vietnam Era Veteran" ---they served in Korea or Germany or Thailand. I tell them "thank you" and "welcome home."

Sure, my war was in Vietnam, but I reflect on how many people it took to sustain me. There's a saying about "in the rear with the gear" or "I was rear echelon." Excuse me, but your job was just as important as mine. It took 10 of you to keep me going. I don't care if you were a truck driver, a unit clerk or an aircraft mechanic—you served. You were subject to the same regulations and orders that I was. Your MOS or duty station could have been changed at any time. There were quite a few times when I wished I could be "in the rear with the gear" and I'm not embarrassed to admit that.

When I got commissioned, I also got a 6-year commitment (two years of active duty and 4-year reserve time). I spent my two years active at Ft. Lewis, training Basic Trainees. We had a poster in my Orderly Room that had a silhouette of a soldier with his ruck and weapon and in the background was a field of white crosses. The statement at the bottom of that poster was "had I the proper training." That was my daily motivation. I could have left Active Duty at that time with no orders to Vietnam, but I decided to stay in to get promoted to Captain (it was called "Voluntary Indefinite") and within 10 days I had orders to Vietnam.

You say you never fired a weapon or pulled a lanyard during "actual" combat, and I say I wished I had never had to do that. It's a feeling you don't forget, and it sits on your shoulders for the rest of your life. I'm sure most VN veterans wished they hadn't.

Your MOS or duty station doesn't define you as a veteran. You stood up, you showed up, you stepped to the line, you raised your right hand, and you took an oath. Be proud of your service and sacrifices no matter where, when, or how you served.

You are no less a veteran. I am proud to call <u>ALL</u> veterans my brothers and sisters.

Home: Back in the World
Carl T. Zipperer

There was a place in my mind called home
It was the place with my family where I had grown

My family and pretty wife were all there
Each day, I closed my eyes and said a prayer

In hopes that I would return to the places
and see all my loved ones' caring faces

I longed for the fields and gardens of home
And all the places where I used to roam

I missed the streams where I used to fish
The place where I could cook my favorite dish

At home, I carried car keys in my pockets
And had no worry about incoming mortars and rockets

A place where I could sleep in peace all night
Without having any drifting flares in sight

No quad-fifties loudly pumping out lead
No worry about sappers while in my bed

Little did I know that when I returned
That many things for which I had yearned

Could be so different than they were before
And some did not even exist anymore

Yes, while I was dodging bullets in flight

My pretty young bride was out in the night

All the hard-earned money sent home was spent
Only my bride and Jodie knew where it went

I never fully returned from that useless thing
As Steve Winwood with *Blind Faith* sang
"And I've done nothing wrong
But I can't find my way home."

The Veterans Healing Farm
Allen Utterback

Veterans Day of 2023 in North Carolina, my wife and I go to the Veterans Healing Farm to see the Traveling Wall. Instead of descending into the hidden memorial, the wall comes up, which I missed, this little ironic foreshadow of the day. As we walk, looking at Veterans, their families, fellow Americans, I see replicas of the Global War on Terrorism monument. A brief thought crossed my mind that my name could have been up there, but then it dawned on me that I wouldn't be holding my wife's hand today. That is when two names from the wall haunt my mind. I don't want to remember, but there is no way to stop it.

I am back in Taji, Iraq on my first tour becoming a combat Veteran. The location is Gunner Gate, and it is my first day on the main gate where you can go in and out to Main Supply Route Tampa. I sit behind a M-2 belt fed 50 caliber anti-aircraft machine gun nicknamed Ma Deuce mounted on a beast of a vehicle called a paladin. The sun is setting where the orange sky and sand become one as a Humvee comes into view. Two more Humvees follow and then I notice the last truck is the only one with a gunner in the turret. I sense that something is seriously wrong, hands on the weapon, and shout to Sgt. Lerch about the vehicles. Distorted sound from the radio and Sgt Lerch is at the gate cable dropping it as the Humvees slalom between blast walls, over speedbumps made from broken tracks, and continue to gain speed. I keep a sharp eye waiting to see if anyone is chasing them through the dust. Nothing. The three trucks speed through making the first right turn to the medic shack. Sgt Lerch latches the cable back to the metal hook on top of the short blast wall and walks over to Spc. Wells and me. "That was the left seat, right seat ride patrol. They were hit with an IED and that is all I know."

He tells Wells to take point here and runs back to the medic shack as the sun finally disappears. My bad gut feeling continues to escalate as our relief shows up. Wells and I hear, "we still got a job to do, now let's do it" as we walk away from our post. The gun trucks are already gone from the medic shack as I listen to the rocks crunching under our boots. A slight distraction from my bad gut feeling. We get back to the guard shack, do guards mount, and hop into the back of the 5 ton to go back to our company area.

We barely have time to drop our gear before getting orders for an immediate formation. I focus up when I hear, "At ease." Now I listen to words I don't want to hear. "Today is a day of loss for our division. The training patrol was hit with an IED, and we had two KIA in our division and one in the unit we are replacing. We will be saying goodbye to our fallen comrades and sending them home to their families." It comes from First Sergeant Smith's lips as he keeps a strong voice. His body language spoke of sadness and anger as we are called to attention to be marched off to the build with battle crosses to say goodbye. I decided in my head in front of the three crosses that failure will not be tolerated and no more of my soldiers will get hurt or die on my watch. Can I back those words up? If I fail, at least I will die trying.

I come back from this memory to my wife talking to me looking worried, but I cover it up quickly. Least that is the lie I can swallow because I am not willing to talk about this with her. I know she knows something is wrong, that is the problem with marrying a smart woman. She drops it for now and we spend the rest of the day together. One day I might talk about this with her or someone who is not a fellow Veteran. One day, but not today.

Death by PTSD
Gerry Nieters

    Matthew Livelsberger's tragic suicide in front of Trump Tower hotel in Las Vegas was a selfless act of mental anguish
    In his cell phone he recorded "I needed to cleanse my mind of the brothers I lost and relieve myself of the burden of lives I took" This screams of PTSD to me! (i.e. Post Traumatic Stress Disorder)
    Why is this tragic outcome all too prevalent in our warriors in the military and domestically in our police and EMS?
    I fully understand his feelings. I was an infantry battalion surgeon in the Vietnam War and have been diagnosed with moderate PTSD I lost six medics out of fourteen in six months. I've visited the Vietnam Memorial (The Wall) in Washington, D.C. 27 times, crying uncontrollably the first 26. I have inner turmoil.
    PTSD is a normal reaction to a very traumatic event or events and is common not only in the military but also in our police and EMS public servants. At the time of the horrific event our emotions or feelings are suppressed so we can carry out the immediate task at hand and not be immobilized or ineffective by the tremendous feelings of horror, disgust, fear, sadness, shock, etc. The mind numbs these feelings so we can perform what immediately has to be done.
    These feelings are not forever forgotten or lessened though. They fester, simmer, percolate and build. They seek expression at a later date. In other words, they grow and are exaggerated with time and always seek expression. This leads to a violent uncontrollable eruption or outpouring of emotions usually in the form of anger or sorrow as an explosive angry outburst or as in my case uncontrollable crying. In both cases these outbursts are totally beyond our control, as are the feared nightmares.
    We can't have that, we must control our emotions. So we stuff them. In stuffing them, we therefore start denying them. By denying them we become withdrawn not only with ourselves but with society (i.e. our family, and friends). We withdraw into ourselves emotionally. We may numb ourselves with drugs or alcohol. We may run from ourselves by doing very adventurous activity to the

point of being self-destructive. We may join a rigid group so we don't have to think but just accept their philosophy. This could be a motorcycle gang or a religious sect.

But most of us simply withdraw into ourselves giving the outward appearance of normalcy while keeping our inward battle to ourselves. In doing so, we withdraw from not only ourselves but from society (i.e. friends & family) as well. We are internally damaged individuals wrapped in a shell of outward normalcy. We never talk about it hoping it will lessen or go away. It never does! What does this mean for us? STRESS. Stress will always seek expression if not mentally then physically. We lead stressful lives, and as a result, so do our loved ones.

So we are in this rat race of denial. We think we're normal. Why can't others see that? We are coping just fine, but we're not.

Our downfall is trust. We have difficulty trusting. We're afraid to trust! For if we can't trust our own emotions or emotional outbursts, how can we trust others. But we try. We truly trust our fellow combatants but have guarded trust in others and our institutions.

When we realize that trust has been misplaced or abused, we are devastated. This leads to further withdrawal or may lead to the ultimate withdrawal- suicide.

May God grant Sergeant Livelsberger the trust, peace and serenity he didn't know in his life.

Prayer to God for those severely affected by PTSD.

"Help me please or I will be destroyed by the inner tempest of my soul

My emotions are beyond my control

My moral compass is askew

My very being is deeply troubled

The waves of anxiety crash against my gunwales

I am adrift in a raging sea of turmoil without direction or purpose

Please God, calm the turbulent waters

Be my rudder and my compass

That I might steer my boat into a sea of tranquility and peace"

My Talisman
Larry Boggs

Most of us want to believe that we control our lives, our destinies. But when it gets down to the nitty gritty, we know we don't. Our lives are more like a moment-to-moment, day-to-day lease that can be terminated at will by fate, god, or the gods, depending on our beliefs. That is why we carry talismans. We carry them no matter how rational we may otherwise be. We hope, more than believe, that they can keep the bogeymen away when we face danger, the unknown, and we fear that it might not turn out well.

We carry them around our neck, in our wallet, in the pocket nearest our heart. We tattoo them on our bodies. We attach them to our clothes, hats, or helmets. They can be given to us by someone we love and who loves us or by a stranger. They can be found on the street, or they can be a family heirloom. They don't even need to be an object. They can, I believe, be something that we hold in our hearts or minds.

They usually are assumed to have some magical, spiritual, mystical, or religious significance. But they can also be something that we view simply as lucky, the proverbial rabbit's foot. Embedded in them is superstition. Something primitive inside of us needs them, and civilization, education, and religion cannot drive that need out of us, especially when something important, like our life, is on the line. A talisman can never be stolen—a stolen object is unlucky and would bring a curse, not protection.

A particularly powerful talisman is an object given by a lover to their lover before he or she goes to war. Remember the poem written and read by our classmate two weeks ago. The locket pressed into his hand secretly by his girlfriend as her father dispatched him from their romp in his car and how he later held it tightly as rockets and mortars rained down around him at Khe Sanh. That locket was his talisman. His poem is a beautiful and powerful memory of it.

Over my lifetime, my most consistent talisman has been a prayer, the Our Father, which I have said most every night before going to bed. This prayer is to me a request for protection. It says, "give us this day our daily bread" and "deliver us from evil." The rest is just homage to his power and our promise to live a life deserving of his protection that is mostly honored in the breach. The cross to me is an embodiment of that prayer and its request for protection.

The talisman with which I began my journey to Vietnam was a simple cross, the Protestant kind, that my mother gave to me the day I left home. I have worn it since whenever I needed to get through a rough spot in life. Most of the time, like all my other mementos of that year, it is in a box in a drawer in my bedroom. But that cross was not my only talisman, nor the most powerful, that I took to Vietnam. That was given to me by a stranger in the San Francisco Airport.

I was in my Army uniform, so was he. He walked up to me and said, "going or coming back?" He knew the answer before I replied "going," as I looked greener than my uniform. I was unmistakably a FNG just out of training. As quickly as I answered, he reached into his pocket, pulled out a small piece of paper, saying, as he gave it to me, "this is for luck." I took it, thanked him, and put in my pocket. After further small talk and after he parted, I looked at it. It was a twenty-five-cent military pay currency. I kept it throughout my tour—even though it was the price of a beer at the enlisted men's club.

When I returned, and I was again at the San Francisco Airport, waiting for my flight home, I was the stranger in uniform that walked up to one of us, who looked as green as I looked a year before. I said, "going or coming?" Before he could answer, I reached into my pocket, pulled out that twenty-five-cent MPC and gave it to him saying "this is for luck." I told him how I got it, and then said, "I hope it brings you as much luck as it brought me." I have prayed often that it did and that he too passed it on when he returned.

Why was that little, nearly worthless piece of paper so powerful in my mind? It was from someone who had gone through what I was about to go through and who knew my fears as he had had them too. We feared being in the field, dying. That stranger, that piece of paper gave me hope that things would turn out ok. He and his little piece of paper said, 'this is not a death sentence, I came back, so can you." And that is why I kept it, held it close and passed it on.

Glossary of Terms

Agent Orange: Agent Orange was an herbicide and defoliant the US used in Vietnam as part of its **herbicidal** warfare. Its name derives from the orange barrels that it was shipped in.

ARVN: Army of the Republic of Vietnam.

C-rations: Individually canned, prepackaged meals.

Chinook: The Boeing CH-47 Chinook is an American twin-engine, tandem-rotor, heavy-lift helicopter.

Cobra: The Bell AH-1 Cobra is a two-bladed, single-engine attack helicopter manufactured by Bell Helicopter.

Concertina Wire: A type of barbed wire or razor wire that is formed in large coils that can be expanded like a concertina.

Claymore: The M18A1 Claymore is a directional anti-personnel mine. The Claymore is command-detonated and directional, meaning it is fired by remote control and shoots a pattern of metal balls into the kill zone like a shotgun.

DMZ: Demilitarized Zone.

FOB: In Iraq, Forward Operating Bases. Bases or outposts or camps.

GWOT: Global War on Terrorism. An international military and diplomatic campaign launched by the United States following the September 11th, 2001, attacks by Al-Qaeda. Its most notable military campaigns were OEF/OIF.

Ho Chi Minh Sandals: A form of Sandals in Vietnam made from recycled tires, distinctive of Vietnamese soldiers. Called "Ho Chi Minh sandals" by Americans.

IED: Improvised Explosive Device, a bomb constructed and deployed in ways other than conventional military action.

KIA: Killed in Action.

Klicks: Kilometers.

LLRP: Long-range reconnaissance patrol, a small, heavily armed reconnaissance team that patrols deep in enemy territory.

LZ: Landing Zone.

M-16: The M-16 rifle, officially designated Rifle, Caliber 5.56 mm, M16, is a United States military adaptation of the Armalite AR-15 rifle. In 1964, the M-16 entered military service and was deployed for jungle warfare in 1965 for Vietnam.

MP: Military Police.

NVA: North Vietnamese Army.

OEF/OIF: Operation Enduring Freedom/Operation Iraqi Freedom. Military campaigns for Afghanistan and Iraq.

OP: Observation Post.

POW: Prisoner of War.

Route TAMPA: One of the major MSRs of Iraq.

MSR: Main Supply Route.

Ma Deuce: A term used to describe the M2 .50 caliber heavy machine gun. Soldiers used the 'M2' portion of its name to create the term: 'Mother of all Machine guns'. This use of two letter 'M' and its 'M2' nomenclature created another term; 'M Deuce', which turned into Ma Deuce.

Humvee: the term used to describe the military's most common light duty and personnel transport vehicle. High Mobility Multipurpose Wheeled Vehicle (HMMWV).

Paladin: a 155mm self-propelled artillery piece whose chassis is like that of a tank and the modern version of the Vietnam era M109.

Left Seat/Right Seat: Used to describe when one unit is outgoing (Left Seat) and another unit is incoming (Right Seat). This allows for the incoming unit to transition into operational control, while being able to learn from the outgoing unit that has operational experience.

IED: Improvised Explosive Device

5-ton: Describes the family of Medium Tactical Vehicles (FMTV) that are the workhorse of military medium duty vehicles.

C-5: a heavy lift, intercontinental-range aircraft, capable of carrying oversized loads and/or a large number of personnel.

CONUS/OCONUS: Contiguous United States/Outside Contiguous United States, describes either the 48 States and D.C. minus or the 48 Staes, D.C, Alaska, Hawaii and the U.S. Territories.

DMZ: A Demilitarized Zone is an area in which military installations, activities, or personnel are forbidden to enter and is used as a boundary between two opposing military forces. Noticeable DMZs include those used in the Vietnam and Korea era conflicts.

Jarhead: a term used to describe members of the U.S. Marine Corps, due to their Dress Blue uniforms and cropped haircuts making them resemble mason jars.

PT: Physical Training.

Battle Buddy: A term used to describe the notion that a Soldier should always be in the physical presence of another Soldier for a variety of safety reasons. The term has grown to include activities off the battlefield, during training, and off-duty, as well as the emotional support of fellow Soldiers.

C-Rats: Used to describe prepackaged canned foods that military personnel used when fresh food is not available. Notoriously infamous for their taste and of dubious origin.

MRE: Meals Ready to Eat. The current name for 'C-rats'.

Slicks: Helicopters used to carry troops or cargo with only self-protection armament.

SOF: Special Operating Forces.

Tet: January holiday, Buddhist lunar New Year. Buddha's birthday.

Tet Offensive: On January 31, 1968, 70,000 North Vietnamese and Viet Cong launched the TET offensive against South Vietnam Villages, cities, and military bases. It was a turning point for the war.

VC: Viet Cong. Officially the National Liberation Front for South Vietnam was an armed communist revolutionary organization in South Vietnam, Laos and Cambodia. It fought under the direction of North Vietnam against the South Vietnamese and United States governments during the Vietnam War, eventually emerging on the winning side. It had both guerrilla and regular army units, as well as a network of cadres who organized peasants in the territory the Viet Cong controlled.

Warrant Officer: A class of Soldiers that are above enlisted and below commissioned officers. They are technical experts in their fields and make up of a very small number of personnel. Besides helicopter pilots, all W.O.s must currently serve in the military with the rank of Sergeant.

HKIA: Hamid Karzai International Airport where the last American soldiers left from.

Contributors

Beth Angel served as a dog handler in the USAF, including an extended tour overseas (Osan AB, ROK). She is a veteran advocate and mother of three and presently works with veterans and their Service Dogs in Training.

Dr. Marsha Lee Baker has shaped herself as an advocate for peace and nonviolence through civic actions, theatrical stories, and educational activities. She is a Professor Emerita at Western Carolina University, having taught rhetoric, composition, and pedagogy. Ongoing collaborative studies include Pace e Bene, the Center for Action and Contemplation, and Quaker Friends. Influential conversations and life with her father and husband led her to write for healing.

Tom Baker is the son of an old "Brown boot" Sergeant Major. He joined the Army at eighteen and went to Vietnam as a Parachute Rigger with the 101st Airborne. After four-and-a-half months as a rigger and part of a QRF, he transferred out to be a door gunner with the 1/9th 1st Cav. He was shot down twice.

Monica Blankenship was a USAF Nurse from 1974-1981.

Larry A. Boggs is a native West Virginian, born in 1945. Drafted into the US Army in August 1968, he was sent to Vietnam in May 1969 as a Field Wireman (36K20), but served as a Personnel Specialist (71H#0) with HHC, USARV Headquarters at Long Binh Post. He took his Law School Admission Test in Vietnam and graduated with a JD from Georgetown University Law School in 1973. Larry now lives in Arlington, Virginia. After years of denial, his body now proclaims with tattoos his pride at being a Vietnam Veteran that had boots on the ground in country even if they were polished most of the time.

Wallace Bohanan is a US Army veteran and served one year in Vietnam from September 1966 to September 1967. He served at Dong Ha, Camp Carroll, and the DMZ with a 175mm Artillery Battalion.

Alan Brett served in the Army from 1966 to 1971. He served in Vietnam from the beginning of 1967 to the end of 1969, and then worked for the Veterans Administration for 34 years. He was the Team Leader in Vet Centers in Casper, Wyoming, and Ft. Collins, Colorado, and the Coordinator for the Posttraumatic Stress Disorders Clinical Team (both inpatient and outpatient programs) at the VA hospital in Perry Point, Maryland. He currently lives near Asheville in North Carolina.

Capt. Donna Culp USAF, BSC (Biomedical Science Corps) (1985 – 1992) U.S. and Germany; CIA Support Officer – Logistics and Administration (2005 – 2013) U.S., Jordan, Iraq, England ; Volunteer in Ukraine working with U.S. and Ukrainian NGOs (Non-Governmental Organizations) (2022 – Present)

Charles R. Duke served with the U.S. Air Force in Southeast Asia and Vietnam as a pilot in AC-119K gunships. After finishing military service as an instructor pilot, he spent several years in corporate work and then thirty years as a university business professor. Retirement is occupied with wood, words, and watercolor.

Dorian Dula is a Marine Corps infantryman wounded in Hue City, Vietnam, during the Tet Offensive of 1968. Later, he received bachelor's and master's degrees and worked the majority of his life with government contractors in the D.C. Beltway.

Emiliano Enea is a first-generation American who lived a stone's throw away from the Twin Towers and was returning home from the Army Recruiter when the Towers were attacked. He served as an Airborne Infantry NCO for multiple OEF/OIF during the GWOT era. Currently, he is focused on helping fellow Veterans adjust to civilian life, just as those Veterans before him helped his transition as well.

Richard Epstein was a 26L20 (Microwave Radio Repair). He served in Thailand in 1966 and went to Vietnam and back to Thailand as a civilian contractor from 1968 to 1970. Richard has hosted a venue for veterans on the National Mall for twenty-five years and co-hosted a poetry workshop series at the Walter Reed National Military Medical Center and at Alfio' Restaurant in Bethesda, M.D. as part of the Veterans Writing Project. Richard has been featured at the Philadelphia Ethical Society, Dog Tag Bakery in Georgetown, the Aspin Hill Library, the Evil Grin reading series, the Mariposa reading series, the Kensington Row Bookshop, the Annapolis Book Store, and the Silver Spring Civic Center. He has produced an anthology of veteran poetry read on the National Mall and has authored two books of poetry about his time in Southeast Asia. His poetry has appeared in Maryland Bards Poetry Review, the Beltway Poetry Quarterly, Poetica, War, Literature and the Arts (USAF Academy); Military Review (Army University Press); the Wrath Bearing Tree; Schuylkill Valley Journal, and others.

Michael D. Hebert was in Vietnam in 1972 on the USS Newport News, then went to Basic UDT/SEAL (BUD/S) training. He was an officer in the Navy SEALs for five years. After leaving the Navy, he served as a CIA Operations Officer for 23 years, retired and served another 15 years as a CIA contractor.

Stephen Henderson was in the United States Marine Corp India Company Three Four, 3rd Marine Division 69-70.

Ironically, **Tom Hickey's** first experiences with war were as a Peace Corps nurse and then as a contractor for the State Department in Central America. After further medical training in the States, he worked as a primary care physician in the US Navy, in a private practice in New Mexico, and finally at the Charles George VA Medical Center in the mountains of Western North Carolina.

Col. John T Hoffman spent the majority of his 31-year military career as a helicopter pilot, a Military Police Officer, and an

Intelligence Officer working in the anti-terrorism, weapons of mass destruction, and force protection areas. He served as a combat helicopter pilot and as an Infantry Platoon leader in the Republic of South Vietnam. After the events of September 11, 2001, he served an appointment with the U.S. Department of Homeland Security. Col. Hoffman is a 1969 graduate of Georgetown University in Washington, DC, and is also a graduate of the US Army Command and General Staff College at Fort Leavenworth, Kansas, and a 1994 graduate of the US Army War College at Carlisle Barracks, Pennsylvania. He is the author of the recently published book about the last year of US Army combat operations in South Vietnam, The Saigon Guns.

Larry Kipp is a Former Dustoff medic and former university biology professor.

Ron Kuebler is a retired Engineer/Speech Pathologist and served as Infantry Sergeant (Intelligence) with D and HHC Companies, 5th/46th Infantry Regiment, 198th Light Infantry Brigade, America Division at FSB Gator near Chu Lai Defense Center from 9/69-8/70. Wife: Margot (since 1973); three Eagle Scout sons: Scott (deceased), Gregory, Matthew. Grandson: Arwyn Scott Kuebler. Currently residing in Flat Rock, North Carolina, United States of America.

John W. Mason served as a Marine Corps rifle platoon (infantry) commander for eight months in Vietnam. He was wounded twice, medivaced back to the States, and "Permanently Retired with Disability" due to multiple shrapnel wounds. Following USMC service, he graduated from law school and practiced law in Asheville for fifty years. He has provided pro bono legal services, creating several non-profit veterans' support organizations, including the North Carolina Veterans' Writing Alliance Foundation, Inc.

Harold (Ted) Minnick was drafted into the Army in 1966, went to Field Artillery OCS and was sent to Vietnam as an Artillery Battery Commander in 1969. He served 8 years Active Duty, 12 years in the NC National Guard and 20 years in the Army Reserve, retiring in 2006 as a Lieutenant Colonel. He worked for the State of North Carolina for 31 years as a Hydrogeologist. He and his wife reside in Black Mountain, North Carolina.

Roy Moore was a Sgt for the United States Air Force from 14 Feb 1984 to 17 Dec 1992. He was a crash firefighter at Howard AFB Panama. He participated in various activities concerning the drug wars in the eighties in Central and South America, cross-trained into a disaster preparedness technician, and was stationed at Pope AFB in Fayetteville. He is from the mountains of Asheville, North Carolina.

Gerry Nieters immediately finished his internship was drafted and sent to Fort Sam Houston for one month's orientation and then sent to Vietnam. He was assigned to surgery at a 400 bed 36th Evacuation Hospital in Vung Tau for six months and was then assigned to the 2nd battalion 39th infantry brigade 9th Division as a battalion surgeon for six months in the Mekong Delta. He DEROS'd Sept. 1968. After residency training at Duke University in Diagnostic Radiology, he practiced radiology in western North Carolina until his retirement in 1998.

Steve "Buck" Owens is a retired Chief Warrant Officer Five (CW5) in the United States Army with over 35 years of service. Steve holds the Legion of Merit and Air Medal, in addition to deployment awards for Kosovo, Afghanistan, and Africa. He flew Attack, Reconnaissance, and Medvac Helicopters, in addition to being the Commander of a C-12 airplane detachment where he flew utility and reconnaissance with Task Force ODIN. He lives in Western North Carolina with his wife, Cathy, of 33 years, and has two grown children, Marissa and Steven.

Allan Perkal served in the US Air Force from 1965-1968, Vietnam 1967-1968, Medic, 26th Casualty Staging Flight. Retired VA PTSD Therapist. Lifetime member of Vietnam Veterans of America. Co-Chair, Vietnam Veterans of America, Healthcare Committee

William "Pete" Ramsey served with the First Infantry Division in 1969-70. His forefathers have all been Infantry since 1775.

Army Sgt. **Sarah Scully** enlisted after 9/11 with a bachelor's degree in journalism so she could offer her writing abilities to tell the Soldiers' story as an Army print journalist from 2003 to 2008. She is now an advocate for treatment for PTSD from Military Sexual Trauma and a Ms. Veteran America 2024 Finalist.

Mike Smith served as an Engineman Third Class (EN3) on US Navy Minesweepers, doing Vietnam (WESPAC) tours aboard USS FORCE MSO-445, USS LOYALTY MSO-457, and USS IMPLICIT MSO-455 in 1969, 1970, and 1971. Patrol assignments for all three ships were at or near the DMZ between North and South Vietnam. Duties included Coastal patrols, fire support for coastal fire bases, board-and-search interdiction of arms smuggled from North Vietnam, and mine sweeping on the CUA Viet River south of Phu Bai and north of Hue City.

Ron Toler was in the 9th Special Operating Squadron, DaNang AB Vietnam 1971-1972. 912th Air Refueling Squadron, Taiwan '72, Thailand '73, Okinawa '73.

Allen Utterback served in the US Army as a Food Service Specialist from 2003 to 2008. He completed two combat tours with 1st Armor Division and the 82nd Airborne in Iraq to include patrols and convoy escort. He grew up in South Carolina and his hobbies include being with family, taking care of animals, reading, writing, and guns.

Jacqueline MacLarty White, originally from Ontario, NY, has resided in Asheville, NC, for over 30 years. US Navy Corpsman, Puget Sound and Bremerton Washington, 1965-1967, Vietnam era.

R. Kevin Wierman USN 1981-1990 & 2001-2008 = Machinist Mate First Class – Nuclear Power Training Unit Instructor - Engineering Laboratory Technician – Engine Room Supervisor – Submarine Warfare – Seabee Combat Warfare – Inshore Boat Unit Engineer & Gunner EXTRAORDINAIRE!

Carl Zipperer served in the US Army from May 1969 to July 1971. He served in Vietnam as a Warrant Officer aviator helicopter pilot from July 7, 1970, to July 7, 1971. He was based at Chu Lai, flying UH-1H Hueys for the American Division, 16th Combat Aviation Group, 14th Combat Aviation Battalion, 176th Assault Helicopter Company.

Dean A. Little, PA-C was a Medic at the 95th Evacuation Hospital at DaNang and was intermittently on loan to the III MAF Cap Team in Goa An, vietnam from 8/69-8/70. Mr. Little was a medical staff member for 30+ years at the Charles George V.A. Medical Center in Asheville,NC. He is married with two adult children and two grandchildren.

www.ingramcontent.com/pod-product-compliance
Lightning Source LLC
Chambersburg PA
CBHW031137090426
42738CB00008B/1119